DIALOG:
Dating & Marriage

by George R. Riemer

CONSULTANTS:
Francis Bakewell, S.J.
Aaron L. Rutledge, Ph.D.
Sidney Callahan
Alfred R. Joyce, M.D., F.A.P.A.

HOLT, RINEHART AND WINSTON, INC.
NEW YORK, TORONTO, LONDON

NIHIL OBSTAT

Richard J. Schuler, Ph.D.
Censor Librorum
October 4, 1967

IMPRIMATUR

✠ Leo Binz
Archdiocese of St. Paul
October 4, 1967

PRINTED IN THE UNITED STATES OF AMERICA

03–064575–1
90123 19 98765432

To my mother, who, a Kansas prairie widow with seven children, advertised for a husband and so met my father.

Author and Consultant Data

GEORGE R. RIEMER, lecturer, editor, and former teacher, has a Licentiate in Philosophy from St. Louis University. He is a member of the Society of Magazine Writers and the Education Writers Association. He has written about intra-family communication for television and national magazines. He produced and edited for magazines Pope John XXIII's JOURNAL OF A SOUL. In DIALOG: DATING & MARRIAGE, Mr. Riemer demonstrates how writing-oriented instruction can serve individual development and education—his central thesis in HOW THEY MURDERED THE SECOND R (W. W. Norton & Co.).

FRANCIS BAKEWELL, S.J. is a teacher in Regis High School, Denver, Colorado. He conducts retreat work throughout the United States and is actively engaged with the Pre-Cana and the Cana Conference.

SIDNEY CALLAHAN: is the author of several books and the co-author of six children. She lectures extensively in the United States on the changing role of women in our time and on sex in the modern world. Her book, THE ILLUSION OF EVE, is translated into five languages.

ALFRED R. JOYCE, M.D., F.A.P.A., is Professor of Psychiatry at Fordham University (School of Social Service) and Director of The Iona Graduate Division of Pastoral Counseling.

AARON L. RUTLEDGE, Ph.D., Head of the Counseling and Psychotherapy Program at The Merrill-Palmer Institute, has twice been President of the American Association of Marriage Counselors and is Past President of the National Council on Family Relations. He is author of PREMARITAL COUNSELING.

Acknowledgments

When I write down the names of all who have helped me in one way or other, I note that I write two names heavier than others and with underlines for emphasis. One is William J. Ward, adminstrative editor of the Catholic Educational Division, and the other is Margaret Flanagan, my secretary.

Before I thank them formally, I want to acknowledge my debt to Reuel L. Howe and Maurice S. Friedman for introducing me to Dr. Martin Buber. I read Reuel Howe's THE MIRACLE OF DIALOGUE while still stunned by the Broadway play, WHO'S AFRAID OF VIRGINIA WOOLF. The book and the play together gave me the idea of applying *I and Thou* to marriage. John L. Thomas, S.J., and Dr. Eleanore Braun Luckey encouraged me to think this idea along. I'm grateful that they did for I respect both their scholarship and their intuitions. Maurice Friedman and Mme. Annette Petite-Veau in different important ways each helped me toward some understanding of Dr. Buber.

I owe thanks to about two thousand anonymous high school students who wrote for me what they thought about dating and marriage. I thank Robert Godfrey, S.M., Rev. William Heffron, Rev. Vernon J. Kuehn, John Teeling, S.J., Sister M. Verona, Rev. John N. Wurm, Sister Anne Marie Hill and Mr. Charles Frier for getting my questions to these students and getting their answers back to me.

I thank Mrs. Audrey Moorman for introducing me to Rev. J. L. Baglio and his Fiat House in Minneapolis, and for loaning her family as test readers. In this context I can also thank Ted and Lory Lothamer of Denver's Cana Conference for insights to the dialog of marriage. I am grateful to Lawrence Tarbell, lay theologian and consultant for Holt, Rinehart and Winston, for his reading of my original manuscript.

Let me thank Ray Robinson, Mina Mulvey and Wade Nichols, editors of Good Housekeeping Magazine, who first encouraged me to write about children and marriage. Their assignments gave me a chance to study the field of interpersonal communication, a field which now totally absorbs me. Their open support and interest in my *written-opinion* research gave me a chance to discover that the freely written statements of students was a neglected but extremely valuable teaching device. This experience helped me formulate the writing-oriented instruction I recommend to teachers who use this book. What my students wrote for me is not at all so important as what their students will write for them.

I'm grateful to Margaret Kelley, project editor, for her conscientious advice and expert copy editing. I am also indebted to Robert Bull for his cover and book design. The graphic artist is a co-communicator with the author. Mr. Bull and I must both wait on our readers to learn whether we came through or not.

All incoming and outgoing words poured through an IBM Selectric driven by Margaret Flanagan. Her cheerfulness and unsighing patience let me see the mirage of only a few days' work separating me from each chapter's end. We need such mirages, all of us, and we need the people who let us believe them.

It was William Ward's idea that I do this book. He saw it complete even before it was started. I'm grateful for his steady, long-distance vision.

Let me thank, finally, all my married and unmarried friends, my happy and unhappy friends whose experiences each in some special way contributed to the growth of my own thinking and, consequently, to this book.

GEORGE R. RIEMER

Table of Contents

CHAPTER ONE: DATING IS A MARRIAGE WORKSHOP 1

1 *Dating Tests Communication* 3 *The Date Was A Bomb* 7 *Does Double-Dating Solve Communication Problems?* 8 *Analysis of a Date in the Making* 22 *Date Identification Guide* 25 *The Single Teenager* 28 *How the Custom of Dating Started*

CHAPTER TWO: BECOMING 39

39 *Man's Central Frustration: Rejection* 43 *The Central Drive: To Be and To Be More* 44 *The Central Sense* 46 *The Self* 53 *Self-Image* 59 *Where Do Self-Images Come From?* 62 *How to Beat B.S.I.*

CHAPTER THREE: DECISION MAKING 65

65 *Decision Making: The Central Means of Personal Becoming* 68 *Little Decisions Are Important* 70 *Guidelines for Decision-Making* 71 *Before Deciding* 75 *After Deciding* 76 *The First Decisions in Life: Why They Mean So Much* 77 *How to Test Your Advisor* 80 *Identification*

CHAPTER FOUR: COMMUNICATION 85

85 *Communication: The Central Means for Social Becoming* 87 *Communication and the Listener* 88 *Communication Failure* 89 *Moodiness* 92 *Lack of Common Interests* 93 *Poor Sense of Humor* 94 *Misunderstandings* 96 *Conceit—Not Listening* 96 *No Place to Go* 97 *Family Interference* 101 *Lack of a Car* 105 *Lack of Money* 106 *Age Difference* 107 *Race Difference* 107 *Lack of Clothes* 108 *Communication: How It Starts* 115 *Martin Buber* 119 *I-Thou* vs. *I-It* 120 *The Problem of Seem-*

ing 121 *The First Principle of Dialog (Openness* vs. *Prejudice)* 126 *Dialog* vs. *Communication* 130 *Barriers to Dialog* 132 *Breaking the Steady Date: A Crisis of Dialog* 134 *Fifteen Dialog Dodges* 140 *Some Rules for Dialog* 143 *Touch Communication* 145 *The Sex-Centric Date* 148 *Keep Out of Steamy Dark Corners* 149 *Four Consequences of the Sex-Centric Date* 152 *Aggressive Necking Is Anti-Dialog* 154 *What Do High School Students Think of Virginity?* 156 *Communication and Sex Instruction*

CHAPTER FIVE: MAN AND WOMAN: DIFFERENCES AND DIALOG 159

159 *Dialog and Difference* 163 *Prejudice & Co.* 167 *Dating Roles* 175 *The Woman* 176 *The Girl's Three Escapes from Home* 177 *The "Weaker" Sex—True?* 180 *Masculine and Feminine* 191 *The Dialog Marriage*

CHAPTER SIX: THE ENEMIES OF BECOMING 197

197 *Why Marriages Go Wrong* 200 *Drinking and Broken Marriage* 205 *Alcohol: Stimulant or Depressant* 208 *Alcohol* vs. *Becoming* 209 *Alcohol and Communication* 212 *Student Attitudes Toward Marrying a Drinker* 216 *Adultery* 217 *The Teeny-Marriage and Adultery* 219 *Five Alleged "Causes" for Adultery* 222 *The Inside Causes of Infidelity* 223 *The Pay Off* 224 *Adultery and the Law* 229 *Why Fidelity?* 231 *Promises* 234 *Irresponsibility* 237 *Conflict of Temperaments* 238 *In-Laws* 240 *Sexual Incompatibility* 245 *Mental Illness* 247 *Religious Differences* 248 *Money Problems*

CHAPTER SEVEN: THE MARRIAGE DECISION 252

252 *Early Marriage* 253 *Why Girls Leave Home to Marry* 256 *Legal Age for Marriage by State* 257 *The Ideal Age for Marriage* 263 *Early Steady-Dating is the Way to a Teeny-Marriage* 267 *Maturity* 270 *The Marriage Decision* 274 *What Facts do We Have to Deal With?* 275 *#1 Legal Contract* 277 *#2 Economic Unit* 277 *#3 Sacrament* 283 *#4 Images as Realities* 285 *#5 Sexual Relationship* 286 *#6 Dialogic Relationship* 290 *#7 Social Unit* 291 *Know Your Feelings* 297 *What is Your Image of Love?* 299 *Ask Advice* 301 *Don't Decide Beyond Your Power to Act* 303 *Images: The Mature Marriage*

Dating Tests Communication

The most popular teenage social game in the United States is dating. Four, six or more can play, but it is primarily a game for two. Partners work toward a common goal but play under different handicaps and are threatened by different penalties. Like other games, it has a variety of serious remote objectives, the most ultimate and serious being marriage. Like other games, *dating* is *exercise.* Skipping rope, dribbling a basketball, and dancing are exercises in physical and rhythmic timing. Dating is important to adult social life because it is an *exercise* in *person-to-person* timing. This timing is needed in communication and decision-making.

Dating is a kind of *fitting* room where we tighten or let out the manners and mannerisms we'll wear in later life. Here we can look at our impressions of ourselves.

Dating is a *testing lab* where we learn to recognize human changes and man-woman differences. We can question what it means to be a mature man or woman and test how we measure up to this meaning.

Dating is a *sales problem,* a *money problem*, a *time problem*, a *transportation problem*, a *planning and organization problem.* It's hard enough for one person to work out these problems for himself alone, but daters have to solve them on behalf of another person, too, and *with* that person. Such team decisions call for a subtle understanding of complex personality differences, of different goals, needs, values and attitudes. The process itself of decision-making isn't as private as when a person decides something for himself alone. Dating decisions have to be explained and accounted for. They must satisfy both partners. They must be attractive to both. Nimble powers of communication, therefore, are needed.

Dating is a *communication workshop* where we can try to express our thoughts and feelings to persons of our own age and development, usually persons as unskilled and inexperienced as we are. At the same time, we are supposed to encourage others to contribute their thoughts and feelings to us.

In this exchange we can learn which of our family-grown ideas can stand broader exposure. We may learn with embarrassing suddenness that we have few interests in common with others.

Can we stand differences of opinion without sulking or fighting? How will we handle a rejection? Will we quit by sinking into a mood or disappearing behind the milky veils of a daydream as we might have done at home? We'll find out in dating.

The hardest test of dating is that it forces us to balance *self* love against *other* love. It matches our willingness to share against our desire to possess without sacrifice and compromise.

The person you choose or refuse to date today at seventeen or eighteen or nineteen may be a person you began liking or disliking thirteen or fifteen or sixteen years ago.

Does a date bring on spasms of cramping anxieties? Constant unreasonable fears of being turned down or passed by? Does your lack of confidence paralyze you in meeting new persons? Are you miserably unhappy and lonely unless you have a steady dating partner?

If spending a few hours with another human being can distract you days ahead of time in school, fill your mind with worries and doubts, make your palms sweat and your face break out in pimples, examine your earliest years of life.

You learned the characteristics of your dating behavior during those years. Whether your pattern consists in talking or kissing or dancing or listening to records or eating, the first traces of that pattern can be found back in the small years between birth and the age of four or five or six. Human behavior—dating behavior included—has its start in the primitive attitudes of infancy and childhood.

The feelings and reactions you repeatedly or vividly experienced in those years shaped your *first habits*. These first habits were strengthened or changed by later experiences.

You might have changed neighborhoods during grade school. You lost old friends and had to make new friends. Your body changed. You discovered sex. You had trouble in school. You had trouble with your parents. You experienced the happiness of love and then, too, the loneliness, the disappointment and agony of love.

You had to cope with these changes. You coped in various ways. Sometimes you solved your problems in the old ways you were used to, other times you modified your old methods to fit the new situation whatever it was.

Every crisis, every shock, every important change affected you somehow. You formed new habits on top of old habits.

Some of the older habits withered away, some became stronger. Even today, this process of weakening or strengthening is still going on in you. But all your decisions and actions can somehow be traced in straight or crooked lines back to your first habitual actions.

Does this mean you are not free? Are you a victim of what you did during childhood when you didn't know better? No, you are not a victim—not unless you let yourself be one.

The patterns of your early life *influence* your present life. You must try to identify and understand this influence. Many of your early patterns are your allies. Be grateful for them. Other patterns handicap you. In either case, the patterns merely influence your decisions today. They don't manage your life.

The Date was a Bomb

In April of 1966, four hundred high school senior boys and five hundred girls in St. Louis, Milwaukee, New York City and Denver wrote papers describing their experiences with "bad dates."

> A bad date is one that was either a bomb from the first, or one that was spoiled by something that happened later. Maybe we don't "hit it off" from the first; or maybe things just go wrong and we don't enjoy it; or he ruins it by trying something or we have a fight.
>
> A bad date is going out with a girl and you are all tightened up and not able to relax . . . You feel on edge and can't trust the person . . . both feel embarrassed . . . ill at ease . . .
>
> Had no where to go, it was raining, no one talked and I got a ticket.
>
> Nobody talks, there is an argument, no car, misunderstanding, all you do not is neck . . . You don't have fun, spend too much money, wish you were out with someone else and don't ask for another date, then you know you're on a bad one.
>
> The whole evening was one of stupid, strained politeness . . . the evening seems long . . . seems to last forever, like it will never end . . . nothing is accomplished. Like you just sit around or argue . . .
>
> My date and I were anxious to date each other but when we finally went out we had hardly anything to talk about; so we kept sort of quiet and the evening was shot pretty much.
>
> I feel unnatural in what I'm doing or saying, everything is forced and I can't wait to get home . . . I feel nervous or tense and the girl feels the same . . . you wish you would have never made the date in the first place.

> The boy seems to be bored to death with you . . . keeps checking the clock . . . I keep looking at the time or find myself wishing I was somewhere else.

> I got fixed up with a girl from——. She was quite good looking and everyone told me she was really sweet. Well, it turned out she was very quiet because she didn't know anything about anything, and couldn't talk or express herself at all. She griped the hell out of me because she sat there and waited for me to talk. So we parked and made out.

> Both can't wait to get home . . . both of us regretted going . . . a complete flop, both bored to death . . . You're both unhappy together, it really isn't doing you much "good."

> Go home with an uneasy feeling about this person . . . All night with nothing to say, nothing to do. One big night of BOREDOM!

Why do dates go bad? How can a meeting which promises to be great fun drag across the clock and end, finally, as a dreadful experience, a night to forget?

Some students complained they were forced into dates by their parents. Blind dates, they observed, rarely worked out; blind date girls often turned out to be ugly or unpersonable; blind date boys proved to be homely, dull or too fresh.

> I didn't want to go out with this boy. It was my mother's idea. He was a friend of my brother's. And that's all we talked about all night—my brother. And I don't get along with my brother.

> . . . a dance was given recently at school. I was asked to take this girl. She turned out to be very quiet saying very little. After about an hour I had exhausted my urge to say something. When I had gotten home I felt that the date had been very awkward and was extremely relieved that it was over.

It's easy to see that blind or forced dates can fail, but how about dates arranged freely and voluntarily by the boys and girls themselves—what makes these dates go bad? Is it the lack of money? Lack of suitable clothes? Not having a car? Not having any place to go? Do parents interfere and spoil dates? Do teachers make trouble? Does awkwardness in talk have anything to do with ruining dates?

How To Have A Bad Date

The loudest, longest, most frequently repeated and urgently pressed complaint from both boys and girls was *talk failure*. A high majority of students in St. Louis, Milwaukee, New York and Denver agreed that the life or death of a date depends on communication.

According to these students, the boy may be rich, well-mannered and may streak around in an Alfa Romeo, but if he can't talk, he's dead. The girl may be Miss Teenage America or Miss Homecoming Float. She may own her own car and dress like a Ford girl in *Seventeen*, but if she can't talk, she's dead too.

> He just sits there and stares. You try to get a conversation started but he just says "yes" or "no."
>
> You sit all night and stare ahead while you're driving because you can't get any response or answer out of the girl. She's a flop in my book.
>
> At the end of the evening, I found myself almost hating her because I couldn't talk to her. You find yourself guessing to try to have a good time.
>
> I dated a boy once who barely talked at all. And when he did, it was dull, dull. That same boy asked me out every weekend for over a year. I never went out with him again after that first night.
>
> I did all or most of the talk (which got me bored) and had to make all the first moves. I felt like his mother not like a date.
>
> Without communication a date is pretty dull and dead. It makes you feel like an enemy not a friend. All you do is sit and get nervous.
>
> . . . she sits there quietly, answers only yes or no, says nothing else. It becomes quite irritable.

Communication failure on dates may take many forms. But the most common bad date noted by students was the person who didn't respond to anything. He or she showed no feelings, no enthusiasm, no interest in anything. Mr. Deadmouth. Miss Emptyeyes.

WHAT GIRLS SAY ABOUT BOYS

You look for someone to talk to. If your date won't talk to you, you feel that he isn't interested enough to say anything to you. You wonder what he's thinking about. After a while you wish you wouldn't have gone and this is a bad date that you don't want repeated.

When I date a guy for the first time, that I don't know very well, there is usually some very heavy silences. This is due largely to a lack of knowledge of each other and you don't want to sit their and pump questions into your date all night.

He can't hold a conversation. Has no manners and doesn't treat me with respect. All he thinks of is to park and make out.

The first date may be frigid, but the ice should be broken by the end of the evening. If you're on your second date or third date with a fellow and it's like this forget it, it's a bad date.

It was just like saying hi and goodbye with a few words in between.

I went out with a fellow once and I mean he was SILENT. He made no effort whatsoever to carry on any conversation. I realize that perhaps he was shy or nervous and I did try to help him along, but to no avail. I had a lousy time and I think he did, too.

. . . approximately ten words all night . . . This one guy I knew would never say anything until you said something first. Then it would take him awhile before he said anything back . . . "yah" or "no" . . . the boy was one of those u-hum guys. I think that was the only word he knew . . . on a blind date he had nothing to say and what he did say wasn't too much . . . I shut up and couldn't get back to my usual talkative self. The date was horrible . . . When he did say something he spoke so softly that I either had to ask him to repeat it, or after awhile I would just nod my head in agreement, not knowing what he said.

WHAT BOYS SAY ABOUT GIRLS

She never does anything, just sits and mopes all evening . . . I don't want her to be cold . . . usually can tell when the conversation moves freely whether she is having fun or not.

Took out this girl and she couldn't talk or converse at all. I mean she wouldn't talk. I asked her sixty-nine thousand questions and she still wouldn't talk. That was the most wasted night in my life.

The girl seemed very impersonal—never wanted to talk about anything, just answered "yes," "no," "I suppose." Didn't want to do anything.

On a blind date at the girl's house everything was all right but the moment we left she didn't say a word except for a few no's, yes's, doesn't bother me, etc.

She wanted to tell me something important but didn't know how . . . so she just kept silent.

Each of us was upset by personal happenings previous to the date and spent the evening staring ahead instead of forgetting the earlier happening and enjoying the evening.

The girl I was with acted like she lost her best friend. I asked her what was wrong, but all she said was "nothing." This shot the whole evening, and I almost didn't walk her to the door. Fortunately I didn't get into an accident on the way home.

We were riding along, no one saying anything. I said something, she said "Uh-huh." It teed me off because she sounded so bored and conceited . . . she didn't laugh at anything and just sat there vegetating. . . . If you have no place to go you just sit around and have small talk but pretty soon you'll run out of small talk and then you'll just sit there . . . I had a date with "sealed lips." I would have enjoyed myself more if I'd been alone . . . She was real quiet, never saying two words. . . . The girl would not talk and when she did it was only one word answers, "yes" or "no." I felt terribly out of place . . .

Does Double-Dating Solve Communication Problems?

Double-dating is fun. It's easier than single dating. There's more to discuss and more to do.

For boys, double-dating acts as an adjustment step between ganging around with the guys and more serious dating. Some boys can't stand the strain of being with a girl alone. They find it easier and more fun to talk in the company of their buddies. A boy may balk at going to a dance alone with a girl but will go if another boy goes too. There are certain practical reasons for doubling: boys who don't have cars can double with boys who do. For girls, double-dating is safer than single dating. It's usually more fun too. But *double-dating may multiply communication hazards.*

We were water-skiing—we doubled with a real cute girl. My steady paid attention to her and I knew he was attracted to her so I was cool. A break down in communication.

My date was more impressed in the other pair than in me, and nothing I could do or say helped. I felt like a fifth wheel.

I doubled with a couple of yoyos who didn't say a word all night and they naturally had a miserable time.

Usually what goes on is the guys talk to each other and the girls talk to each other. That's double dating. Even if one of the guys sits in back it goes that way.

. . . you try to make conversation, and he just gives a sharp, crisp answer and doesn't try to make you feel at ease. He directs all his questions and statements to the other girl completely forgetting about you.

Communication *is* the date. The person I was doubling with is a real card. He is fast with words and ideas. This took most of the attention of my girl as well as his. When I couldn't compete with him I felt sort of low.

If a person has a strong personality a date with others would be fun. But if the person isn't too outgoing, maybe, a few single dates would "break the ice."

Sophomore year I tripled. It was the first time I dated this girl. The whole night was spent talking with the others. There was no more than a few words spoken between us. We didn't know how to talk to each other. We never did go out after that.

The boy carried on the conversation with everyone else in the car but me. As though I wasn't there.

One date for me was spoiled when the couple we doubled with, rather the girl did *all* the talking. She didn't stop from the time we left until we came home again. All she talked about was all the boys who are crazy about her. Very *disgusting!*

Which do you think is more fun?

	Boys	Girls
Double-dating	53	74
Dating alone	43	22

Purdue Opinion Panel 1959

Analysis of A Date in the Making

Arranging a date is a game in which the ball moves from boy to girl to boy and back to girl, faster and faster until there's a score or the play is called dead. It is a match of tensions. Both boys and girls fear they might not be wanted. The girl is afraid no boy will ask her out. The boy is afraid the girl will turn him down.

At the start, the boy wants to meet a girl. The girl wants to meet a boy. He wears a clean shirt and he combs his hair—if that is his style. She puts on lipstick, makes up her eyes. She stands in the "right" places. She wears the "right" clothes, tight or floppy, short or long. Her hair is closely cropped or teased up full, or long and silky.

Neither boy nor girl starts with anyone specific in mind. They are influenced by ideals, by taste and past experiences but their interest is open, general and not exactly defined.

While dating is a balance of rejective and acceptive powers, in some ways the system seems weighted against the girl.

1) The girl's dating period is shorter since her chances for marriage diminish with age. Statistics say that at fifteen, a girl has a 90 percent chance of marrying sooner or later; at thirty, she has a 25 percent chance; at forty-five, her chance has dwindled to 10 percent.

Before the age of twenty-two, a girl's likelihood of getting married is greater than a boy's. Between twenty-two and forty-five, the boy's chances of getting married are greater. After forty-five, both male and female have about a one-in-ten chance.

2) Girls outnumber boys so that competitive feelings develop among girls.

3) The girl carries the knowledge that she is biologically adapted to bear children and is therefore expected to bear a special responsibility. The girl is penalized much more severely for her moral failures than the boy is for his.

4) A lot of the girl's time, money and attention must be given to cosmetics and fashion since her looks first attract a boy. She feels she must accent her sexual features to attract boys. Her capacity for companionship and her ability to communicate may be passed by.

5) Girls usually have to wait to be asked. They don't have equal opportunity for initiating a date. Usually the only move they can make is to be where the boys are. In most communities, girls are still likely to be marked as too eager or too aggressive if they make the first move.

6) After a first date a boy may simply not call again. If he isn't interested in the girl, he may have no honest reason for calling. The girl often feels this is unfair, especially if she likes the boy. She would like to know why she was rejected, but it is considered bad form for her to call the boy to find out.

There is no need to feel sorry for the girl. Usually she's well able to cope with her handicap. Marriage has been jokingly defined as *a man chasing a woman until she catches him.* This can be said of dating, too.

A successful date is a good looking boy that I've been trying to get.
Girl, 18, Milwaukee

Must the boy always be the one to start a date rolling? Is it always out of place for the girl to make the first move?

A girl may start the play, but she must make her move according to the conventions of her social group. She may start conversation with a boy who interests her, but will she suggest going out together? If she proposes a date, what will she propose? A house or beach party or dance may be all right, but will she hurt her reputation if she suggests a drive-in movie?

Two general norms might guide the girl. 1) She ought not take the initiative where a date will burden or embarrass the boy in some way. For example, she shouldn't suggest a drive if she knows the boy will have trouble getting a car. She shouldn't suggest an expensive film or play for which the boy is expected to pay. 2) Unless she expects an evening of heavy necking, she ought not tease a boy by suggesting a date at her house, for example, when her parents are away.

Round #1 is usually the Boy's Move

He sees her. He has seen her before – around school or at the games or in his neighborhood or at a dance. He likes or dislikes the way she looks. He likes her sound, her laugh, or he can't stand it.

Here is his first decision: he will go up to her or he will put it off and turn away. He is acutely aware that if he bids he may be rejected. This would embarrass and confuse him. He thinks of ways to build up his self-confidence. The girl is usually unaware of this struggle. If the boy rejects her, she is usually unaware of having been rejected, since the rejection took place in his mind.

Dating is a balance of veto powers. The first veto is customarily the boy's. He decides not to exercise the veto and aproaches her.

WHAT MAKES A BOY MOVE TO MEET A GIRL?

. . . are they good looking and friendly. That does for starters.

The thing that attracts me first is her looks. If she's good looking I then start talking to her.

First a physical attraction because when I first see a person I wouldn't know what type of personality she has. Talking helps fill out the personality chart which is the most important aspect in lining up a date. Things in common help.

Looks and figure, if I don't know her. Personality and looks and figure if I do know her.

Good looks make a boy look at a girl, but personality should be considered. If a girl is ugly most boys will not look at her, no matter what type of personality she has.

Is she cute, is she intelligent, is she capable of having a good time?

Physical appearances for the first date. If her personality is bad there will not be a second date that is for sure.

Physical attraction helps, but usually does not decide my date. A girl who takes good care of herself by being clean, keeps her proper weight and dresses well attracts my attention.

The first round and first veto is necessarily superficial. It is based mostly on *appearances*. Four appearances are involved: *face, body, dress* or *clothing style*, and *behavior*.

To me a real girl is one who is quiet, is clean and a neat, decent dresser. Someone without a lot of makeup or high frilly hair. Someone between fifteen and eighteen.

. . . in most cases the person just looks for a good looking girl. This should not be the first thing to look for, but it helps.

She must be at least a little pretty with a halfway decent personality.

The way in which the person appears to handle herself in public.

The way a person speaks and acts, also the way she dresses.

The way in which she conducts herself. I feel for one thing she must have a sense of humor, and not too much of "an air of superiority" toward others. Before I even suggest dating a girl I "investigate before I invest."

Tone of voice and reactions to things said.

If I know the person to an extent, it is personality, how she carries on conversation, her sense of humor. If I don't hardly know the girl, much will depend on how she looks.

A very beautiful girl may discourage some boys, particularly those who lack self-confidence. They feel at once they could never compete with a more handsome boy, or they are afraid a beautiful girl has to be expensively entertained to be held.

A girl who dresses in poor taste often attracts only persons with poor taste. Many boys will gape and whistle at a girl who looks "sexy" and who dresses provocatively, but they would feel uncomfortable going out with such a girl. The boy who likes his date to look like a "sex cart" may have some neurotic need to prove he's really masculine.

WHAT MAKES A BOY REJECT A GIRL?

A girl with bad legs, bad breath, bad body, bad mug, and bad personality.

Poor body, dull personality . . . .

An ugly fat girl with a bad personality.

An ugly thing that makes you feel like being by yourself.

Personality counts more than anything, however, I wouldn't go out with an ugly babe, no matter how good her personality is.

Round #2 is usually the Girl's Move

She sees the boy's approach and may veto immediately by refusing to meet him. She turns away or ignores him while he talks. She may even ask him to go away. If she decides to meet him, she smiles and accepts his greeting. If her hand goes to her hair to see if it's in place, it's a sign of special interest.

WHAT MAKES A GIRL DECIDE TO MEET A BOY?

This will depend on how sure the girl is of herself, what she's interested in, her experience, her values, needs and goals. According to student reports, the boy's approach is the first thing that affects her decision. Some girls decide on very little else than front. Some girls know better and wait to see what's behind.

A good looking boy with a good build and personality, with a car and money.

The externals, being the only thing which is evident, are what I judge. His appearance, personality and manners.

That the person be fairly settled down and sincere. A warm smile often moves me.

A person who is neatly dressed and groomed. A boy that acts like a gentleman. Someone who's friendly and has a good sense of humor not someone who's "a stiff shirt."

How he treats me and also, his moral standards. I consider how he looks, if he's nice looking instead of looking hoody.

If a person is neat and cares about himself he will take care to see that you are treated right. Clothes can tell you much about a person. If he is dressed in the "in" clothes he is a bandwagoner. I want an individual.

APPROACH

The way the boy approaches you. His appearance. If he is calling cause it is a blind date I would accept maybe out of curiosity or if I have been told about him ahead of time it would help.

Their personality and approach. His looks are a minor thing

PERSONALITY

Personality. If a boy is nice and his personality is attractive and he isn't physically repulsive, I'll go out with him.

I love a person who has always a friendly smile. I don't think looks count as much as having a good time with the person who is almost just like you and has the same interests as you.

I am attracted to a nice personality and a good sense of humor. The boy needn't be Cary Grant, but he must be a nice boy with a good reputation—not conceited or standoffish.

HE GETS ALONG WITH OTHERS

How he acts in front of a group of girls—if he has good manners, if he makes a nice appearance.

The personality and the way he handles a situation. I always watch the way he acts toward children; if he goes along with them and doesn't make fun of them, he's got to be fun. I like a talker, because I'm not.

The person's sense of humor, sincerity, interest in others and ability to get along with others.

I like a clean-cut fellow, one who I know I won't have to distrust. A good sense of humor when it's necessary. I don't believe in fellows laughing at the wrong time.

HOW HE TALKS

If he is a gentleman, polite, clean-cut, doesn't use bad language.

The way he treats a girl, his language, his looks but most of all his personality.

If they are real friendly and talk to you a lot. A boy who curses all the time for only little things is someone to keep away from. He has no control.

WHAT KIND OF BOY DOES A GIRL REJECT?

. . . sloppy, irresponsible, insincere, rude and doesn't have any plan in mind.

. . . acts vulgar, or very rowdy and insists on taking you someplace or doing something you don't want to do.

. . . someone who is immature when he is in mixed company. Ex: loud, giddy, vulgar, etc.

. . . some boys don't think about how the girl feels. They use any kind of language around girls.

. . . a person who curses a lot and trys to show off.

. . . a dull talker or a boaster and a selfish character.

Boys that are very nervous and not sure of themselves make me nervous too.

Round #3: What Moves the Boy to Move?

He talks and hears her answer. He may like her personality at once and begin to think how he will ask for a date. If the boy senses he may be rejected, he won't risk asking. He has to feel encouraged.

The way she reacts to me. If she takes an interest in me.

If the person seems to like me, and if the person is attractive and charming. She mustn't be loud or forward.

Her attitude toward me (in other words, if she notices me when I speak and if she is pleasant when speaking to me).

Her looks, her personality (especially) and her attitude toward me. If I don't think the girl would be compatible and friendly, why bother taking her out.

SHE MIXES

. . . must have a sense of humor and show a kindness toward everyone.

The person's personality, the way she acts around people and how well we get along together.

If she acts nice around me and everyone else.

. . . friendly, good sense of humor, easy to talk to and fun to be with.

HER REPUTATION

What the person looks like, the person's personality, and what other people think about the person.

THINGS IN COMMON

If she is cute, clean-cut and honest. She must have a good personality and also if she is interested in the same things you are.

If she is like me in any ways, such as: personality—humor.

The person should have a similar personality to yours to begin with; your interest should be the same.

PERSONALITY, LOOKS, IDEAS ABOUT THINGS

I first look for a person who is outgoing and charitable. If they have these two qualities they usually are also happy, well-rounded individuals. Good looks help, but personality and temperament come first.

The attitude of the person plays a big part. If she is the type that is individualistic, has her own ideas, is out to have a good time in clean fun and is understanding if certain circumstances arise where I come late, etc.

A person who has a congenial personality, who shows me that I could have a good time with her, and help her to have a good time.

An appealing and charming personality.

The way a girl acts according to being a person who has a complete personality.

Mostly personality because this is what counts in life.

The person must be someone you know you can have a good time with. I think she should have some good looks too.

. . . good personality, good looking, not necessarily popular, and someone I feel comfortable with.

If a girl is popular I notice her. If she's doing things I can watch her and like her because we all know she's there. I never watch girls who are too quiet.

. . . the girl's personality, how cute she is and her shape. How well she can carry on a conversation and how she acts, her maturity.

On the other hand, the boy may find her personality unattractive. Veto possibilities pop up in rapid variety like targets at a carnival gun-counter. The girl speaks with a harsh accent which he decides is unpleasant and cheap. Or she may be too easy, too willing and he suspects there must be something wrong with her. Or she may be aggressive and he's had enough of bossy women at home. Or she's too smart for him and makes him feel uncomfortable. Or she doesn't answer fast enough and he decides she's dull. Or she answers even before he's finished talking and he thinks she's giddy-headed or too nervous. If the boy is primarily interested in the girl's body, he may still move to date her even if her personality is unattractive and in spite of any or all of these defects. But if he wants more from the girl than her body, he'll cancel the idea of going out with her.

The way a person looks and what you think you will get away with.

Round #4 is the Girl's Move

For one reason or other the boy has asked for a date and the veto is now with the girl. She must decide why he asked. She must decide whether he is attractive to her. She doesn't like his approach; it's uncouth, inconsiderate, too smooth, too sharp, too defensive, phoney, crude, too frightened. His conversation is dull. He brags. His speech is monotonous or vulgar or uncultivated or effeminate. He makes suggestive remarks. All he does is talk about his car. His nails are dirty. She may turn him down for any of these reasons and others too subtle to define, but what makes a girl say yes? According to the girls' statements, the boy's *personality* is the deciding factor.

Whether or not the person is nice—will talk, has manners, seems to want everyone to have a good time.

The way a person really is, his personality, not the clothes he wears or the money or car he has, only what he is and what he feels.

A nice personality not a smart aleck or show off.

. . . the boy has to be considerate and thoughtful, level-headed and polite. It helps if he's good looking, but it's not necessary. I'd prefer that he didn't smoke or drink.

If he has a lot of confidence, if he is good looking, if he is intelligent, if he is a good talker.

Looks, of course help to a degree, but don't always prove dependable. I would rather go out with someone maybe not so cute and have a great time.

Most important is the personality. 1) I can get along with this personality. 2) Although looks shouldn't be the single factor, it does play an important part, believe it or not.

I went with a boy who wasn't quite my type . . . the friends he hung around with when we went out weren't the kind I liked or enjoyed being with.

Does he like people and get along with them?

. . . able to get along with others even if they aren't in the same crowd or age bracket.

LOOKS

I like a good, clean-cut guy. If he's really sweet and has a good personality and is easy to talk to, he is my idea of an ideal mate. looks aren't that important.

. . . looks are what attract me first, but if the guy doesn't have morals or doesn't respect me, to me he's really pretty ugly. Personality counts most.

The first thing that helps me to accept a date is his appearance. A person doesn't have to be real cute or handsome. I believe that if someone has a nice personality and is clean cut it will show through his skin.

I had a movie date with a fellow. When he came to the door he had baggy-striped green pants on, a maroon colored jacket, a yellow pinned striped shirt and red tie with burnt orange shoes. Talk about embarrassing to go to a downtown show! And, no sense of humor!

LOOKS AREN'T EVERYTHING

I look for sincerity, a sense of humor, honesty and integrity, good-grooming, a generous nature. Looks and money and race and religion are, I think, only incidental. Maturity attracts me also.

I guess a lot of kids just go by looks, but I think that personality matters a lot. I think I'd rather have a lot of fun than just sit there all night and look at something pretty.

His personality is the main factor. Sometimes though I just don't feel like going out with a particular boy. For no reason just a feeling. The second factor is if anybody knows him and where he can be found in his spare time.

If he looks like a good kid and fun to be with I would probably accept.

UNCONCEITED

. . . sounds sincere. If he acts as if he is doing me a favor—I won't accept.

NO PANSY

His personality and some looks. He can't be a pansie that gives in to my every plea and whim. He has to be a mature boy with some responsibility. Has to be somewhat loguacious.

HUMOR

I like to go out with a person who can make you laugh even when you're down in the dumps.

A lot depends on the personality of the person. He must be able to carry on a conversation and be very humorous. He has to be a responsible person.

ACTIVE

If he is friendly, likes a date with a lot of activity. He should like to smile.

DOES HE MIX WELL?

The way he acts in public, the way he dresses, the places he takes you.

SOME COMMON INTERESTS

One that has the same background as you. Has the same opinions on life and religion. Has same interests.

A neat appearance does not denote good looks just careful about his clothing, etc. Also he must have the same principles and basic ideals.

A CAR

If he's good looking, and if I can talk to him. A car helps.

HIS REPUTATION

If he is not a reputable wolf, I will go out with him.

He looks neat and has a fair reputation.

Whether I like the personality of a boy. If he is not a show-off, neat, at least looks half-decent, I'd go out with him. I don't listen to stories others make up. The only way to find out is to go out with him.

RESPECT

One who respects me and acts as if he is proud that I am his date.

A clean looking person, one who displays good manners and thoughtful consideration. A person who respects womanhood and takes a girl for a person, not a body.

He should at least be clean of mouth and clean in his dress. This is for the first date, it usually will go from bad to worse.

. . . friendly and a lot of fun to be with. A boy should also ask a girl to a nice place on the first date, not a drive-in.

COMMUNICABLE

The personality of the person, if he is considerate, honest and behaves like a gentleman and if he can be respected. I like a person who is not a phony or brags a lot.

His personality. Whether he is faintly intelligent, courteous and whether or not I feel comfortable talking to the person.

A person must have an outgoing personality. I must be able to talk freely to him. Looks help at first, but after the first date, they don't really matter.

His personality—a person you can be yourself with. His dress-clothes don't have to be expensive, but he must dress in nice, clean and ironed clothes. Be a nice person.

DECISIVE

. . . sincere, responsible and knows exactly what we'll do that evening.

In a boy I listen to what he says. I listen to his past experiences. How they affected him and how he took them.

Round #5 is the Date Itself

The veto is balanced. Sometime during the course of the date, either boy or girl may decide another date is desirable.

The boy will know whether he's going to call again. He may even begin getting up nerve to ask during the date itself. The girl knows whether she will accept.

What makes a boy ask for a second date is a successful first date. *What do boys call a successful date?*

Mostly I'm attracted by physical appearances on the first date. But the second date is suggested on the grounds of their behavior on the first date.

A girl which you feel at ease with. You find a lot to talk about and think of good things to do. You agree on many subjects and enjoy each other's company.

The good, the true, and the beautiful body and mind. Not necessarily in that order, but some of each. They are what decide me for the second time.

The way they make the first impression on me. Their ways of showing gratitude, and happiness.

I'll ask again if they talk and just don't sit there all night and say nothing. Should not be a show off and don't smoke.

The person's attitude attracts me most to ask again. Good looks help, but aren't necessary.

. . . her character, and if she is truthful or not.

Her sincerity toward going out with me and her character is a good help.

If they are interested in people and they like to dance and are interested in sports. If they have a knowledge of God at least enough to discuss this subject.

Mainly the personality and a neat appearance. She doesn't have to be exceptionally cute for me to take her out again.

When the girl accepts the next date without hesitation or thought.

A girl with good legs. A girl who is fun to talk to and fun to be with.

. . . both parties have a good time and some affection is shown.

. . . A successful date is with a girl who is cute and I made out. A good date is when you go to a drive-in and don't see nothing. When I have a good time and get a lot of lovin'.

When the girl shows you some consideration, a little affection but not going overboard. A kiss would help to make the evening complete. If she accepts a date from you for next week this also completes a date successfully.

A date who can laugh at things funny.

If when you like the girl she likes you and you had a good time.

A good date is one that you can say was worth spending your money on.

WHAT DOES A GIRL CONSIDER A SUCCESSFUL DATE?

A successful date is when you can truthfully lean over to your date, give him a kiss, and say "I've had a wonderful time," and really mean it.

One where I can relax and enjoy myself and feel at ease. One where I don't feel uncomfortable about silences, or I don't have to worry about screaming for help. I like to go out with someone I've known ahead of time.

When you forget about any problems you may have and really enjoy the date and don't mind going home because you know he'll call again.

I consider a date successful if I come home with a feeling of contentment. A feeling that I have had a good night on the town. One where the boy doesn't ignore you.

When nothing embarrassing happens, but even if it does it's all right if you are with the right person.

If he gives you a lot of compliments and makes you feel as though he's glad he took you out.

. . . both feel comfortable around each other.

One who would never make a girl embarrassed to be with him.

A successful date is one which makes me feel good from the beginning of the evening until I come home. One where the boy is a success. I like him and I know he likes me.

. . . he says he'll call you and you know he will. You don't have to hope.

If I come home with a feeling of happiness and seem to have had fun—this is successful. If I can trust my date—it is successful.

. . . he took you someplace nice, was funny, had a good time, kissed . . .

When you have something to do and keep it on light friendly terms.

When a boy and girl have a complete understanding of each other and their moral capacities.

One in which partners go someplace definite—not to park or ride around—talk and laugh, and kiss good-night. Not necessarily one kiss, but no heavy necking.

If I go out and have a good time and know that my date has had a good time also I feel good, if we got through the night without a lot of making out, etc.

I don't think there is such a thing as a bad date. It may be a little boring or it may even be degrading if the guy gets fresh but I wouldn't call it a bad date.

Good wholesome fun shared with close friends is a "successful" date because what you are looking for is some way to have a lot of fun.

When a couple can have a good time through clean fun. Such as bowling or going to eat or a game of miniature golf.

Going to a nice place like a show, dance, etc., going to eat afterwards and then going straight home.

One that can have as much fun alone as one couple as he can have with a group. That you can go somewhere with and have fun even if you just go to get a soda.

When both parties enjoy themselves and have no problems with conversation, type of entertainment, the company you both keep.

If the fellow can make the evening in some way special.

When you both have fun and enjoy each other's company, without having to be on your guard.

So what if the party was a flop or the person was bad, the two people make a date. If you can joke around, laugh at yourself and others without hurting or being hurt.

Dating changes in quality as the daters mature and gain experience with different partners in different dating situations. It starts as an awkward experiment in social relations and ends in the most serious decision most of us will ever make: the decision to marry or not to marry a specific person. It proceeds in stages from group and random dating purely for laughs and good times. Eventually it settles down and becomes *one person's search for one person.*

Date Identification Guide:

A CATALOGUE OF BAD DATES, SOME BOYS, SOME GIRLS, REVEALED BY THOSE WHO'VE HAD TO SUFFER THEM.

THE PUBLIC IMAGE DATER

She feels her best dating college boys or lettermen. A boy with a car is better than a boy without one. A boy with his own car is better than a boy who uses his father's, and a boy with his own car *and* money is still better. A boy with his own sports car is best, but a boy with an Alfa Romeo is better than one with an MG.

He dates only the prettiest girls. The girl who's pretty and goes to an out-of-town school and drives her own sports car is a good catch. A beauty contest winner rates high too.

Some boys and girls aren't very successful on dates but like the reputation that they are successful. They like to excite the candid envy of their friends; they boast of conquests which have taken place only in their minds. They never misjudge a date and always get what they want from it—if you believe them.

LIKES: other Image Daters, Pageant Daters and Car Nuts. It doesn't matter whom they date; it's the talk about it later that counts.

CAN'T STAND: Body Daters, Panic Daters. Image Daters don't really enjoy dates, but they don't expect to since they date for impression and prestige.

DOCUMENT:

. . . he just wanted to go because his friends were going and he wanted to be in the swing of things and I had a terrible time because we have nothing in common.

A bad date would be going out with someone you don't like just to be going out. You don't like him or her so why even bother. You'll sit at a play bored just to tell other kids you had a date.

You go out with someone just to be able to say you went somewhere. I don't think it is worth it.

THE PAGEANT DATER

Pageant Daters prefer events, big parties, shows, roller rinks, games—any place where there's a crowd. This is not merely a person who seeks company to keep out of trouble. The Pageant Dater feels uncomfortable being alone with a date because he or she is afraid of personal communication and commitment.

LIKES: Pageant Daters and Image Daters because they don't ask for anything more than appearances.

CAN'T STAND: Body Daters, Panic Daters and Exploiters.

DOCUMENT:

A boy takes you to some expensive place just to show off. It usually doesn't impress the girl but ends being a bad date. I like it when you can go some place and just be yourself. I'd rather go to an informal show and out for a pizza or go-karting, miniature golfing or just walking down by the lake—any place where I can be myself than go to some formal place.

You have to put on an act like you are having a good time but down inside you wish you had stayed at home.

In order to make an impression while they're talking the couple will talk about anything even if it is stupid.

THE PANIC DATER

This person, usually a girl, is desperately afraid of being left out of anything, thinks it's the end of the world if a dateless weekend comes up. She'll date ANYBODY just to get out of the house. She never drops anyone. The Panic Dater is usually good for a neck night. In a few years time, expect the Panic Dater to become a strategist, full of plots and sub-plots aimed toward getting married.

CAN'T STAND: Exploiters.

DOCUMENT:

You go out with someone just for the sake of going out. Especially when you really don't like the person at all and each seems to only tolerate the other.

THE BODY DATER

He would rather kiss than do anything. Touch, not talk, is his medium for communication. He can't stop rubbing, nudging, squeezing, stroking.

LIKES: Any body that's warm.

CAN'T STAND: Pageant Daters.

DOCUMENT:

See the Sex-centric Date.

THE EXPLOITER

This is often a boy, a strategist full of plots and sub-plots aimed at getting as much as he can from a girl. What makes the Exploiter different from the Body Dater is that he camouflages his early maneuvers. He has a good appearance, good manners, good conversation. He is considerate, unselfish. His evenings are planned and entertaining. He impresses the girl's parents well. When finally the girl is relaxed he makes his move.

LIKES: the Image Dater.

CAN'T STAND: the Pageant Dater.

DOCUMENT:

. . . a boy that talks all night about his previous dates and how he got what he wanted from them . . .

. . . tries to get advances and tells about other dates that were more fun.

THE TRADER

This is a Body Dater who expects affection as his natural right. Three girls describe him:

> A guy who can't talk gives you a bad time and then thinks he deserves a reward for taking you out . . .
>
> All the boy does is sit, doesn't utter a word, but when it's time to leave he expects something in return . . .
>
> The boy expects a reward after a nice evening out . . .

THE CAR NUT

Usually a boy. His real date is the car not the girl he's with. When he's not physically in a car, he's there mentally. He talks cars and smells of gasoline. After being in love with his car all evening, he suddenly remembers his human date and turns into a Body Dater.

The Car Nut isn't aware there are different kinds of girls. He's vaguely aware that he sometimes has fun and sometimes doesn't but he doesn't believe in analyzing his personality or anyone else's. He'd rather take a motor apart. Besides Car Nuts, there are Ski Nuts, Muscle Nuts, Surf Nuts.

LIKES: Panic Daters because they're very accommodating. Image Daters because they appreciate a good car. Pageant Daters because they don't require him to talk about anything human or personal.

DOCUMENT:

> I once went out with a car nut. I know little about cars and I cared less about duel exhausts and sticky pistons.
>
> One person I knew just talked about his car or hi-fi constantly and was not interested in what I had to say. There was no interest on my part though I tried to look interested and absolutely none on his part.

The Single Teenager

Some foolish persons and some lucky persons have married the first person they ever dated. Some persons date and date and never marry. Some start dating late and don't date at all as teenagers. All-boy and all-girl high schools have a high percentage of students who for a variety of reasons never date.

> I don't know what a bad or successful date is. I have only been on one date.
>
> Girl, 18, New York

> I have never been out on a date. I like girls but am not interested in dates. It's a waste of money. I'm probably missing a good time and hindering my character development. I don't have the time and couldn't dance even if I did go.
>
> Boy, 18, Denver

According to a Purdue Opinion Panel poll conducted in 1962, 27 percent of the high school students questioned said they *rarely* or *never* went on dates; 34 percent dated *less than once a week*; 34 percent dated *once or more each week.*

Why do some teenagers never date? There are several reasons:

LITTLE OPPORTUNITY

The fact that girls usually have to wait to be asked keeps many of them from dating, just as the fact that the boy must usually do the asking may keep both boy and girl from dating. Some boys don't know how to ask and are afraid of being turned down. The possibility of being rejected frightens both boy and girl. Some overcome the fright and are better for overcoming it. But some are defeated by it.

NO CAR

Lack of a car may keep some boys from dating because of the distance they live from desirable recreation places. Many boys imagine that girls regard a car as essential and feel that without a car they don't stand a chance of being accepted.

NO INTEREST

Some persons simply feel no interest in dating and resist all pressure put on them by their peers to date. Many date without realizing what they're doing or what dating's all about. They date because it seems to be the thing to do. They date for competitive reasons. They spend many dates wishing they were home watching TV. They act bored, sullen, moody, quiet; they come home bored and unhappy after making someone else bored and unhappy too.

CAREER BUILDING

High standards required for getting into college today may keep both boys and girls at their books. A boy feels awkward going out once a year, so he doesn't bother going at all. He looks on dating as an expensive interruption to his work. Students who plan to enter some form of religious life may date to test their vocations but some may simply feel they're sure of what they want to do and don't need to try themselves out.

A boy or girl may work after school and have to study at night. Some keep busy with school activities and have to use all extra time to keep up their grades.

LAZINESS

An effort has to be made to date successfully. Some students won't make this effort. They won't fix their hair or their clothes. It's all too much for them. Some complain that cosmetics are too expensive, but Eileen Ford, a beauty authority, says that, "the simplest and best cosmetic is cleanliness."

SEXUAL DIFFICULTY

Some don't date because they feel attracted to persons of their own sex rather than toward persons of the opposite sex. Some girls may avoid dating out of abnormal fear of sexual exploitation or from a normal wish to avoid sexual involvement.

PARENTS

While dating is common in the United States, it doesn't mean that all parents have accepted it. Some persons don't date because their parents refuse to let them date. Some have to work to help their parents and have neither time nor money for dating.

UNATTRACTIVE

The feeling that they're unattractive keeps many persons from dating. They may suffer from physical abnormality or may have been ravaged by a crippling disease. They may be handicapped by a speech defect or disorder. The handicap may be temporary, as, for example, teeth braces or acne. Frequently, these persons suffer from oppressive self-images acquired early in life.

How the Custom of Dating Started

Dating starts as a phase in social growth but is refined as a system of decision-making. Between its start and its crucial period of selecting companionship for life, dating is tried by rejections, disillusionments and expensive waste of time. It is flagged ahead by fun, interesting times and talk, and a hunger for sympathy and laughter. It is stimulated by sexual forces and the need for affection. It is driven by loneliness and by friction at home.

Dating is accepted so widely and thoroughly in the United States that we take it for granted. We think teenagers all over the world date as we do. This isn't true. We think dating must be as old at least as marriage itself, maybe older. This isn't true either. It is still a young custom, unsettled and in the process of change.

Dr. David Mace, a famous marriage scholar, distinguishes four patterns of selecting a marriage partner:

1. *Selection by the parents—the young people themselves not consulted.* [This method, once strictly and universally followed in the Orient, is now changing to methods 2 and 3 below.]
2. *Selection by the parents, but the young people are consulted.*
3. *Selection by the young people, but parental approval necessary.*
4. *Selection by the young people—the parents not consulted.*

Each of these four methods of selecting a marriage partner has its own special effect on dating. Parents who have everything to say about marriage will also supervise and direct pre-marriage meetings and associations. This is reasonable for pre-marital associations lead to marriage.

Dr. Mace travelled broadly in the East studying marriage customs. He reports the following conversation in India:

> "Wouldn't you like to be free to choose your own marriage partners, like the young people in the West?"
>
> "Oh no!" several voices replied in chorus.
>
> Taken aback, we searched their faces.
>
> "Why not?"
>
> "For one thing," said one of them, "doesn't it put the girl in a very humiliating position?"
>
> "Humiliating? In what way?"

"Well, doesn't it mean that she has to try to look pretty and call attention to herself, and attract a boy, to be sure she'll get married?"

"Well, perhaps so."

"And if she doesn't want to do that, or if she feels it's undignified, wouldn't that mean she mightn't get a husband?"

"Yes, that's possible."

"So a girl who is shy and doesn't push herself forward might not be able to get married. Does that happen?"

"Sometimes it does."

"Well, surely that's humiliating. It makes getting married a sort of competition in which the girls are fighting each other for the boys. And it encourages a girl to pretend she's better than she really is. She can't relax and be herself. She has to make a good impression to get a boy, and then she has to go on making a good impression to get him to marry her."

Before we could think of an answer to this unexpected line of argument, another girl broke in.

"In our system, you see," she explained, "we girls don't have to worry at all. We know we'll get married. When we are old enough, our parents will find a suitable boy, and everything will be arranged. We don't have to enter into competition with each other."

"Besides," said a third girl, "how would we be able to judge the character of a boy we met and got friendly with? We are young and inexperienced. Our parents are older and wiser, and they aren't as easily deceived as we would be. I'd far rather have my parents choose for me. It's so important that the man I marry should be the right one. I could so easily make a mistake if I had to find him for myself."

Another girl had her hand stretched out eagerly.

"But does the girl really have any choice in the West?" she said. "From what I've read, it seems that the boy does all the choosing. All the girl can do is to say yes or no. She can't go up to a boy and say 'I like you. Will you marry me?' can she?"

We admitted that this was not the done thing.

"So," she went on eagerly, "when you talk about men and women being equal in the West, it isn't true. When our parents are looking for a husband for us, they don't have to wait until some boy takes it into his head to ask for us. They just find out what families are looking

THE ARRANGEMENTS

1890-1910 **In this period, people were very mature in their attitudes toward arranging dates. Mainly because the people who did the arranging were the Parents.**

THE MARRIAGE

1890-1910 A good percentage of marriages made during this period didn't work out . . . for obvious reasons.

for wives for their sons, and see whether one of the boys would be suitable. Then, if his family agree that it would be a good match, they arrange it together."

David and Vera Mace,
Marriage East and West
Doubleday & Co., Inc., N.Y.

Not very long ago marriages and dates in this country were arranged and supervised by parents.

Parents decided who was eligible and who was off-limits. They decided how long the "visits"—dates—could last and how far they could go. The boy "called on" a girl at her home. He visited her there while her father sat nearby reading his paper. The boy could take her to a church social or school picnic but had to specify where he was taking her and guarantee to be home before dark. After a number of meetings, his intentions could be challenged by the girl's father: was he thinking of marrying the girl? If the young man committed himself to marriage, he courted the girl properly.

Calling and courtship, of course, were only for young men old enough to work and support a wife. Younger boys might carry a girl's books home from school, but they didn't go "calling."

People in those days married for different reasons than they do today. Their needs for marrying were different. They had a different *marriage image.*

Some people married for practical reasons: to pool resources, protect property, merge family fortunes, to preserve the family name or tradition. Some married for moral reasons—to keep out of trouble or because they had already got into trouble. Girls married early because what else could they do?

The American way of marriage made an important shift in the 1920's. World War I which brought about the partitioning of Germany and Austria also, unexpectedly, helped partition the U.S. family.

Before the War, the father of the U.S. family was its undisputed, unchallenged leader. He owned a farm or a shop or he worked for someone else who did. He was a carpenter, a mechanic or a plumber. He worked hard as did everyone else in the family.

Each household was a laundry, a bakery, a cannery, a restaurant, a sewing center, a clinic. Mother was in charge of this household. She put her children to work as soon as they had muscle enough to pick, sack and carry potatoes, haul water or ashes. Her own hands were rough and calloused from the rooting, hacking, lifting and dragging work she had to do.

What was the teenager doing fifty years ago? You can be sure he wasn't dating. There was too much work to be done. He didn't have time. He was ball and chained to a variety of unpaid part-time jobs he called "chores." He didn't move much either. He lived on a little piece of ground: the family property, his parish, his neighborhood.

Everyone put in a twelve to fourteen-hour work day just like father. There was one magic difference that made father boss. His work was the only work that brought home *money* With money the family could buy tools, seeds, a wagon—new things. It could order things from a catalogue from Chicago or New York. Mother's work, however valuable and important for the family, wasn't useful to outsiders so it couldn't bring in money. The teenager's work didn't bring in money either.

World War I changed this system. American soldiers had seen different ways of living in Europe. They had become disillusioned, cynical and impatient toward their own prewar ways. Now troop ship after troop ship began bringing them back home. There was a general restlessness, a state of mind looking for change. They wanted to stay in the city. A popular song of the day caught their mood: "How Ya Gonna Keep 'Em Down On the Farm After They've Seen Paree?"

Big city conveniences cost money and father seldom earned enough in the factory to afford them all. Eventually mother went to work, too. When mother started earning her own money, father's days as the solitary head of the house were numbered.

American women won the right to vote in 1920. There was no great rush to the polls, but women were now equal citizens with men. Their new civic equality inspired other changes. They began to study medicine and law. They were accepted in colleges which before had taken only men. Commercial canneries, bakeries, and laundries competed successfully with home-made products. They gave women time and freedom. They used this time to work outside the house.

Americans looking for some firm intellectual authority to respect turned to science. A number of biologists and anthropologists during these years were suggesting man was really little more than a fancy animal and that man's moral codes were nothing but folk superstitions. Sigmund Freud's ideas fell among these attitudes. Freud described sex as the dominant powerful drive behind everything man did. Until this time, sex was a subject family people whispered about out of hearing range of children. Now it became fashionable to discuss the "science of sex" boldly in public. Magazines began to publish "True Confessions." Excited by this new freedom people

SEX + LIQUOR + CAR = FUN–this was the formula of an extremely small group of college-age men and women who rebelled against established courtship manners in the 1920's. John Held, Jr., whose drawing is shown below, partly invented and partly depicted their adventures. The "Roaring Twenties" or the "Jazz Age," as this decade in American history was called, died with the stock market crash of 1929, the year on Held's license plate.

COURTESY MARGARET J. HELD

JOHN HELD, JR.

stretched Freud's ideas beyond limits he himself intended. Catholics reacted to this distortion by ignoring or condemning Freud.

Teenagers changed certain values too. A little over forty years ago, the Purdue Opinion Panel questioned twelve thousand public high school students about their ethical beliefs and values. In 1923, the second *Worst Practice* in the opinion of teenagers was "Sexual Misbehaving." "Killing or Murdering" was #1. "Dancing" was #16.

In 1954, Purdue repeated the poll. "Killing or Murdering" was renamed #1. "Sexual Misbehaving" had fallen to become the third *Worst Practice.* "Using or Selling Narcotics," not even mentioned in 1923, had now taken the second position. "Reckless Driving" was added to the list and "Dancing" fell off the list entirely.

In 1965, "Sexual Misbehaving" dropped to #4 after (#1) "Killing or Murdering," (#2) "Narcotics," and (#3) "Stealing."

One product which strongly influenced man-woman relationships was the low-priced, mass-produced car, particularly the closed car. In 1919, barely more than 10 percent of all cars produced and sold in the United States were closed. In 1924, 43 percent and, in 1927, 83 percent of the cars made were closed cars.

The closed car was a gasoline-driven, one room roll-away apartment. It could be parked out near a country field or on a dark road. It was a "get-away car" to escape from watchful parents and society's guardians.

The social custom of dating as we know it today was invented in the 1920's by a small, rebellious set of American young men and women who defied the courtship conventions of the day.

The first daters were rebels proclaiming and demonstrating ideas of free love and free sex. Girls who dated were mourned by their parents and considered "damned" and "lost." By and by, less defiant, more conservative young men and women also began making their own plans to meet and spend time alone together without asking anyone's approval or permission. Gradually, as more responsible young persons began "dating," the custom gained respect and acceptance.

Only in the United States and in countries recently influenced by the United States is a boy so free to meet and share the company of a girl. The boy must ask only one permission—he has to ask the girl herself.

Six things continue to influence U.S. dating customs making them different from pre-marriage customs of any place in the world:

(1) Mobility—teenagers have wheels.

(2) Money—teenagers have jobs. Many keep the money they earn. European students are usually dependent on their parents for date money.

(3) Co-ed schools—high schools imitate colleges.

(4) More girls can plan on going to *college*. Studies postpone marriage for awhile.

(5) There is more and more interfaith dating.

(6) Matching a rise in interracial marriage, there is a rise in interracial dating.

NEARLY THREE OUT OF FIVE WOMEN WORKERS ARE MARRIED

Fifty-seven percent of all women in the labor force in March, 1964 were married (husband present), and 23 percent were single.

This is a remarkable change from 1940, when only 30 percent of all women workers were married (husband present) and 48 percent were single.

Quiz

1. How old is the custom of dating as we know it in the United States?
2. Name five conditions which make the U.S. dating customs different from those of most other parts of the world.
3. Name four factors which changed the system of boy-girl relationships after 1915.
4. What does dating have to do with marriage?

EXPLAIN THE FOLLOWING STATEMENTS:

5. "Dating is an *exercise*." (An exercise of what?)
6. "Dating is a *test-lab*." (What does dating test?)
7. "Dating is a communication workshop."
8. What period of our life must we examine to understand our dating behavior?

NUMBER OF MARRIED WOMEN IN THE LABOR FORCE HAS GROWN RAPIDLY

(WOMEN IN THE CIVILIAN LABOR FORCE, BY MARITAL STATUS, MARCH 1940-64)

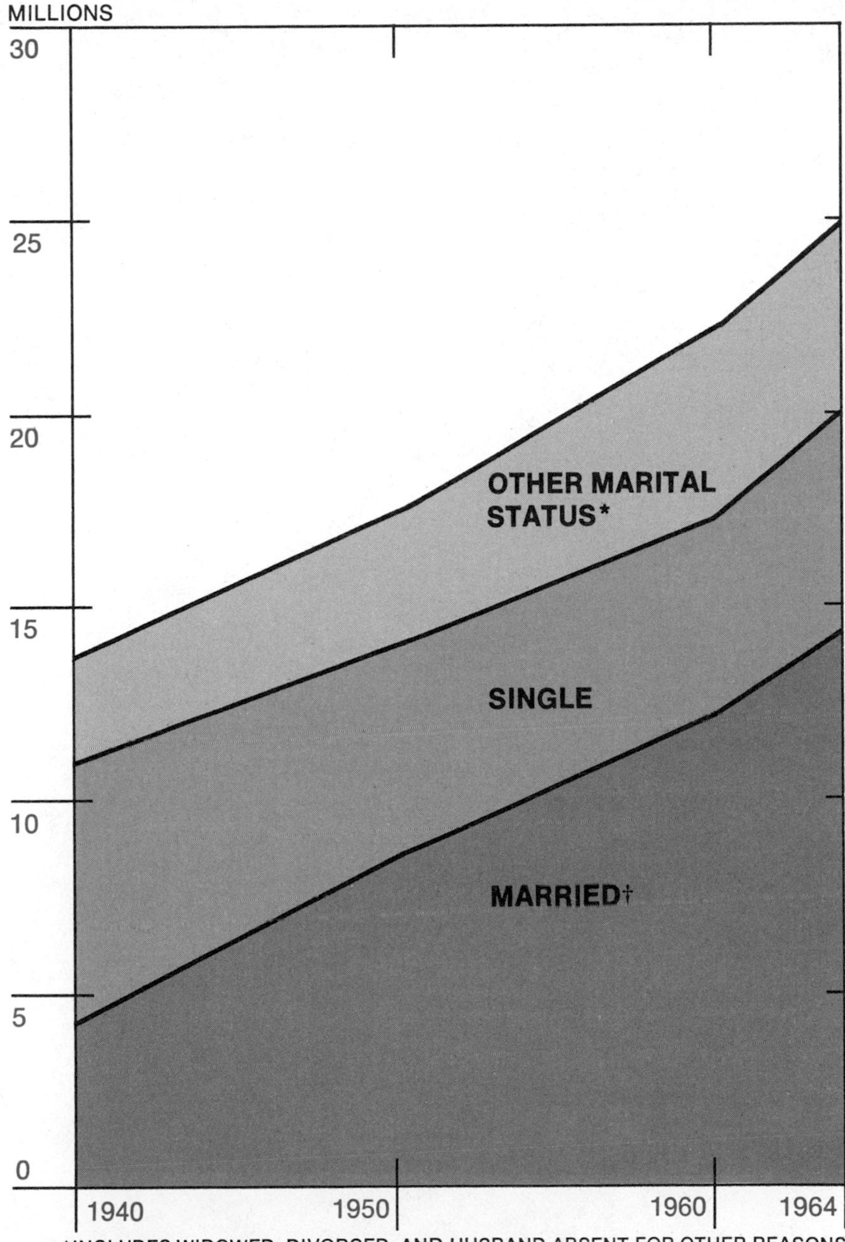

*INCLUDES WIDOWED, DIVORCED, AND HUSBAND ABSENT FOR OTHER REASONS
†HUSBAND PRESENT

Sources: U.S. Department of Labor, Bureau of Labor Statistics; U.S. Department of Commerce, Bureau of the Census.

CHAPTER TWO: BECOMING

Man's Central Frustration: Rejection

You find you have little in common to talk about with your date. You feel everything you say or do is wrong. Your date is a person who has too high of moral standards. You feel un-at-ease.

She acts as if she is bored and doesn't give a damn what you do . . . the girl doesn't realize you're with her . . . the other person seems to have little interest in you, doesn't talk to you or seems indifferent.

A bad date is one which you don't care about the other party, whether you hurt his or her feelings.

A bad date is one in which the girl has been ungrateful for an evening of fun and enjoyment . . . A cold one. The girl who doesn't give a person a good time . . . acts as if she is bored and doesn't give a damn what you do.

We were on our way to the show and no one said a word. Maybe it was because I really didn't like the guy. And he felt it.

A girl may want a certain boy to take her to a dance but he asks someone else. She's been rejected. She wants a certain boy to call her. She stays at home waiting but he never calls. He has rejected her. A boy wants to date a girl but can't even get himself to say his wish out loud to her for fear of being rejected. Or he does ask and his fear is realized—she turns him down.

Rejection is the cruelest hurt a human being can suffer. It is a partial death.

Rejection is a hammer shot against man's most central drive. To understand our fears of rejection, therefore, we have to identify that drive.

Why is it so important to be accepted? Why do we feel such terror about rejection? Our earliest fears of rejection occured in infancy.

The most violent upset of a child's basic sense of good feeling comes from the feeling of not being wanted, not being cared for, not being loved. This feeling can be caused in a variety of ways ranging from desertion and abandonment to subtle changes of voice or expression.

The mother makes her child feel secure by consistent, sympathetic care. She pays attention to his needs, holds him, rocks him, sings to him. She keeps close physical contact with him. Her sense of hearing is never more acute than now. As he grows older, the child needs his mother less and less, though she still has to be there to comfort and reassure him when necessary.

The normal growing child gets protection, food and unconditional love from his mother. She gives him time. She gives him attention. She watches his progress as he learns to crawl. She cheers his first words, shows him off to friends, and she comes when he cries and tries to understand what he means by crying. She encourages him to experiment, to attempt standing and later to attempt walking. She corrects and punishes him, but the infant never seriously doubts for long her continuing love.

The child needs this consistency. Living in the same house on the same street can help the child feel secure, whereas frequent surprise changes of clothing, scents, sound, food and living arrangements may make him nervous.

Families of U.S. Air Force personnel must make frequent moves. A pilot may be based first in Oklahoma, then in Germany, then in Liberia and finally Tokyo, all before his children even finish grade school. Studies of children in such families show they can quickly and easily adapt to changes and are not especially troubled by feelings of insecurity. On the other hand, such feelings may trouble adolescents who have never been moved in their life time. The consistency which the child needs is consistent love. The secure child will have the courage to go out and meet the world. He will be able to accept the uncertainties of the future without abnormal fear and anxiety. He will be able to overcome small waves of anxiety and learn how to master conflicts. How he handles himself in each personal crises will determine the quality of his emotional growth.

ANXIETY: A feeling of dreadful or fearsome threat. Usually of such a vague nature that the person cannot correctly identify its specific source. It is different from fear, which is the emotional response to a consciously recognized threat or danger. Anxiety and fear are accompanied by similar physiologic changes. Anxiety may be so severe that it interferes with effectiveness in living. It may destroy emotional comfort. It may prevent one from achieving desired goals, or satisfaction. When anxiety is so severe, psychiatrists say it is *pathological.*

Parents are quick to deny that they ever reject their children, but in utter truth many would have to admit some feelings of regret since a child's birth raises a storm of inconveniences.

Before birth, the mother had only to eat properly and be careful not to fall. Her body did the rest. But at birth, her relationship with the child changes. Aside from certain helpful instincts, she has to make efforts to care for him. A new baby is work. A child is work.

Parents can't help but resent the cost and sacrifice of time, fashion and style of living that children force them to make. The kitchen seems forever filled with unwashed diapers and soaking milk bottles. The Thunderbird smells of baby.

The baby's first need is the *unconditional love* of his mother. What "unconditional" means is that *the mother must love freely without seeking to get something back for her loving.* The mother who is emotionally mature is ready to give this kind of love. But if she herself is still psychologically a baby, she will compete with her child. If she's never learned the habit of hanging on, of patience, she will resent her child's endless cries for attention. If she feels her baby should provide her with love and affection as a pet would, she's not ready for motherhood. She can damage her child's life by training him to smother her with affection.

Every young mother feels at times that the *da da mama moomee ma* language of her child is a death sentence to her own intellectual growth. She may worry about holding the interest of her husband. Before marriage she may have had a good office job and now misses it. She misses the company of adults. She misses talking about things which interest her. These needs may influence her to neglect her child while she goes out in search of interest and affection herself.

Some parents reject their children by expecting too much of them. A father who is an athlete may expect his son to be tops in sports. He may find it hard to conceal his disappointment when his son prefers reading to throwing a baseball. A father who is a musician may be disgusted at his child's poor show of talent. A scholar may feel his child is stupid. Such disappointments are rejections. In every case, the father fails to see the son as he really is. The child's limitations and differences must be recognized and accepted. Anything less than that is rejection.

An immature mother feels she has better things to do than stand by and watch her baby's attempts to walk or talk. It's quicker to pick him up and carry him. It's safer and less bother to tie the child to his chair or playpen rather than have to watch him. She can't stand watching his struggle with problems. If he's frustrated by an obstacle and cries, she stops his crying by removing the obstacle. She doesn't like the mess he makes when he feeds himself, so she feeds him. Again and again and again, she interferes with the child's natural initiative and spoils his chances of learning to solve problems. She pretends she is doing this to help the child but actually she is rejecting the child and protecting her own convenience.

Rather than let the child take all her time and energy, the immature mother buys toys, sends her child out to play, sits him in front of television. She delegates her child to baby sitters and phonograph records and as soon as possible to school, camp and books.

No matter how the immature mother may try to disguise her rejection, the child feels and knows he is being rejected. He feels his mother is too busy for him. When a child is rejected, he suffers from a sense of worthlessness; he doubts his value in life. He may tend to withdraw or he may try to attract his mother and win her love by being as perfect as he can. He thinks: if I'm good, maybe she'll stay with me. Some children who are rejected become hostile and aggressive. They smash toys, tear books, damage anything prized by the parents.

WITHDRAWAL: a retreat from people or the world of reality. In psychiatry it is a pathological condition seen often in schizophrenic subjects.

WITHDRAWAL SYMPTOMS: the physical and mental effects of drug addicts or habituates when deprived of their drug.

Rejections may take many forms. Parents may reject their children in order to improve their own personalities or advance their careers. For example, a woman may be out of the house a lot doing volunteer work for charitable organizations or political campaigns. She regrets neglecting her children but insists she needs these outside activities or she would be an unhappy wife and mother and, consequently, no good to her family. A man may have a childless jealousy of the attention his wife must give to their children. She placates him, convincing herself her children will learn self-reliance by being on their own.

The act of rejection is the result of a decision and therefore influenced by the rules of decision-making. Frequently, parents who reject their children are violating the first rule of decision-making. Their rejection is a decision to avoid responsibility.

Chief Justice G. Joseph Tauro of the Massachusetts Superior Court, in deciding that the use of marijuana was illegal in his state, wrote this opinion:

> The drug has a great attraction for young men and women of college age or less during their formative years, when they should be gaining the education and experiences upon which to build their future lives. The use of the drug allows them to avoid the resolution of their underlying problems rather than to confront them realistically. While marijuana is not physically addictive in the sense that heroin is, it can cause psychological dependencies. . . .
>
> Its users may not be driven to its repeated use by a physical craving, but they may come to resort to it habitually in order to compensate for real or imagined inadequacies or to avoid real or imagined problems.

The Central Drive: To Be and To Be More

Every infant is delivered into the world with certain drives or urges wound tight as springs. These must begin uncoiling immediately at birth. They will continue to uncoil little by little all through life. He has a drive *to be* and a drive *to be more; to survive* and *to assert himself; to grow* and *become more* and more; *to preserve* himself and *express* himself.

How does the human person *become more*? He has two central means for growth. (1) He *becomes* individually and personally by acts of his judgment or decisions. (2) He *becomes* socially through communication.

The human infant is born with a potential capacity for judging and a potential capacity for communication but these capacities first have to be drawn out and developed by exercise before they can be of any value to him.

How are these two means for *becoming* developed? The human infant develops them by imitation and identification. The child needs freedom to move but is afraid of venturing out too far alone. He needs to feel someone is calling him out, waiting and hoping for him, someone to be near to. He also needs to feel assured that someone strong is behind him backing him up in case of trouble.

These particular needs of attraction and support are met by different persons who appear at different stages of development, helping his coiled drives to unwind, supplying love-warmth that encourages them to open while controlling and directing their spiral expression.

Since the human person's central drive is *to be* and *to be more,* his central frustration will be a frustration of that drive. Growth is stopped instantly and absolutely when the mind cannot judge. Growth is thwarted or sluggish when a person cannot communicate.

But growth doesn't always depend on our own actions. When we act toward others, we have to be received. Our growth partially depends on them.

We *become in others,* when we are recognized, understood, hated, disliked, avoided, ignored and rejected.

The Central Sense

Someone asks: "How are you?" I think *How am I? I have no broken bones, no headaches, no big worry, no special sadness.* "I'm all right," I say. "Fine." I am referring to my BSGF, my *Basic Sense* of *Good Feeling.*

The BSGF is like the bubble in a spirit level used by engineers and disc jockies. When the level is properly balanced, the bubble rides a line in the exact center of the level. When the level is off balance, the bubble glides off center, to one side or the other. The growth of medical science can be traced in the work of doctors learning to define more and more precisely where this balance has been disturbed. We would know very little about physical or mental good health if it were not for sickness. Sickness is the disturbing sign we are somehow, somewhere off-center.

The human being's BSGF is first established by his mother's body before birth. The infant in his womb-envelope becomes used to a regular flow of food and a dependable warmth. The normal heart beat and emotional balance of his mother establishes his own normal emotional state. Her different emotional tensions are communicated when the central balance is disturbed by changes of pressure and pulse and as her hormones pass into his blood stream to affect his endocrine glands. The *feeling of nearness to someone else who also has feelings becomes an important part of his BSGF.*

Then, suddenly, this BSGF is smashed by the violent act of birth. Scientists calculate a baby is shoved out of the womb by a 100-pound propulsive force. He is pushed and pulled out into a factory of wild, hard, sharp sounds. His skin is exposed to a shocking 20° drop in temperature from the 98°F of his mother's body to room temperature. He feels winds and drafts for the first time, some of them blown against his body by drafts of 'gale' strength. He has been used to darkness, now his eyes are blasted by light. He moves from a wet world to a dry one. Formerly, he felt sound with his whole body, now he hears it. To feel his mother's feelings as he did before, he must reach out and touch her.

He cries. His mother answers. What he cries for is the BSGF so roughly and unexpectedly disturbed by birth. His mother answers by protecting him from the cold, raging air. She adapts her hands and arms, her neck and shoulder and the curve of her body to hold him. Skin against her skin, he feels the familiar relaxed beat of her heart. All is well again.

His mother introduces him to food. She makes sounds, low and warm, to assure him that the strange whirling clangor around him isn't harmful. With her help he gets used to light and dark, hot and cold, soft and rough. He gets used to movement and change.

While restoring some of his original sense of good feeling, his mother at once begins to help him acquire a new BSGF, one with *nearness as part* of it *but modified now by his new separate condition.* For the first time, he sleeps separate and alone, feeling now only his own heart beat. Birth has made him a free, individual person. But he is still dependent. Independence is his next goal. It will be his goal for life. He can never totally achieve it except in the vision of God. But even when he achieves some relative independence, he will sometimes experience a dim, obscure hunger. Is it the hunger for God in his future or could it be only the memory flicking over his earliest central sense of good feeling?

THE SECOND BIRTH

Adolescence is a kind of second birth. In the first birth, the infant passes from mother's womb to mother's care. The second birth begins with puberty and ends when the young adult leaves his family home or environment to go out into the world.

Adolescence is called a "second birth" because the adolescent is forced to deal with some of the same crises which he experienced as a newly born infant. He must undergo another ordeal of change. He must find a new basic sense of good feeling.

He feels a force urging him toward independence and rebellion. He feels a balancing force pulling toward conformity. He feels a strong need for personal approval by others, especially by his peers. He suffers if he is not accepted socially. He has strong feelings of inferiority. He has a tremendous need and desire to be accepted by his parents *as he is,* not as they would like him to be. He is being tumbled into a fast, hard knocking, competitive adult world where very few will go far out of their way to help him. He will have to rely on himself.

When the adolescent is ready to break emotional ties with his family, he begins to feel special tensions and anxieties which serve to get him ready to break away. As his body changes, his feelings are in a turmoil. His old self-image doesn't fit his new SELF.

Each physical and biochemical change in the adolescent boy and girl drags along its own network of psychological changes. Family and friends, of course, notice the changes. Either they're pleased or disturbed by them; either they like them or can't stand them. Their attitudes and comments have their effects on the adolescent, too.

This is the young teenager's world: changes inside, changes outside. They have to cope with both.

The teenager's ability to revise his central sense of direction and to cope with change, with loneliness and with love will be as good or as bad as his ability to communicate. Inability to communicate at this time will make the adjustment more painful and cause it to drag on longer than necessary. With some, the adjustment doesn't happen at all. Many marriage conflicts are brought on by unresolved adolescent problems.

The Self

When we speak of *becoming,* we mean the *becoming* of *SELF*. What do we mean by *SELF*?

The SELF has as many different definitions as we have ways of looking at it. We will talk about two, the Actual and the Potential SELF, but it's important to remember that there's only one living, growing, real-life SELF.

A psychiatrist, Dr. Harry Stack Sullivan, calls the SELF *that part of the personality which has alertness, the part which notices what goes on.* It is the person's *psycho*logical activities as distinguished from his *physio*logical activities. It is also called Actual SELF.

Dr. Karin Horney distinguished the Actual SELF from another aspect of SELF she calls *real.* The real SELF is the *source of the energy in each individual which can be directed toward constructive and healthy growth.*

Dr. Erich Fromm has a different term for this aspect of the SELF. He calls it the *true* SELF. The true SELF is that *potential* of the *individual which might be developed given the most favorable social environment.*

In our book, generally, we will use the term *Potential SELF* instead of *"real"* or *"true."* We don't want the words "real" and "true" to be mistaken to mean "actual." *The Potential SELF is the SELF's undeveloped, unexploited power which can become Actual if the person co-operates with certain favorable creative conditions.*

How do the Actual and Potential SELF differ? Suppose you write a bad exam. Your I.Q. indicates you could have written an "A" paper. The I.Q. marks your Potential SELF. Your exam grade theoretically marks your Actual SELF at the moment of writing.

If we say a person has a great capacity for work, we refer to his Potential SELF. If we have a *talent* for music or writing or sports, the talent unused is our Potential SELF. It is our Actual talent when we use it.

What SELF do we refer to when we "appeal to a person's *better self*?" His Potential SELF—the good person he *can* be if he wants.

How can anyone love a newborn infant? It is noisy, messy and utterly useless. It cannot make conversation. It makes only trouble. It is true the infant's Actual SELF is extremely limited. But it has a rich Potential SELF. Much of our love for the infant is directed toward what he can be.

Definitions give a false image of the SELF just as high-speed photography gives a false picture of the moving object. A camera may catch a pitcher winding up, his head turned toward first base. We look at the photo and wonder: how he can throw to the plate with his eye on first? We know he doesn't throw while looking at first. A movie camera would prove that the "still" photo's image was "frozen" out of context. Movie film would show the pitcher's complete sequence of motions. His glance toward first base would then be a single frame in the sequence.

The Actual SELF is never still. It is forever becoming, forever in change. In real life, the Actual SELF and the Potential SELF are one and the same SELF. Both refer to the same SELF in different ways. We separate them only in definitions.

ACTUAL SELF: the total psycho-physical being at any given moment.

POTENTIAL SELF: all of the invididual's powers, capacities, resources *which could become actual* if given attention, encouraged and developed. It includes those talents, powers and capacities which lie waiting to be expressed, waiting to be released.

The Potential SELF is the person *as he could be.*

The Actual SELF is the person *as he is.*

Actual refers to *being*. Potential refers to *becoming.*

THE SELF AND BECOMING

We grow and *become more* generally in two ways: by *impressions* and by *expressions.*

I gain impressions through all my senses, from reading, from honest listening and honest asking. I gain impressions by studying a sunset or staring into a garbage wagon, by a cross-country hike or a walk through the supermarket, by simply feeling good on a nice day.

Impressions are influences – that is, *in-flowings.* They are *imports* supplying my personality.

But I can also *become* more by changing things *outside* my SELF. I can *export* my thoughts and feelings out to others: by talking, by acting and dancing, by the clothes I wear and the way I drive my car, by walking over fresh snow, by writing my name on the side of a building or after a seventeen-word poem or at the bottom of a test paper, by burning a tire mark on the road, by playing a guitar or an iron grating, by telling a joke, making someone yawn, by breaking a window, chalking L-O-V-E in a heart on an asphalt side street, by hitting a Cub Scout, by helping someone off the bus or drawing a mustache on a bus ad.

Expressions are *e*ffluences, or *out-flowings.* They are acts I do. They are *exports* from me to the world outside me.

IMPRESSIONS HELP BECOMING

Impressions are outside things which affect me somehow, change me slightly, urge me to judge them, press me to make some little or big decisions about them. As I judge them and decide about them, I *become* them and I *become more*.

Impressions affect both my Actual and Potential SELF. They engage my Actual SELF. They call out parts of my Potential SELF. Sometimes they challenge parts I didn't even know were there.

Has anyone ever asked you a question which you answered being surprised yourself you knew the answer? The answer was in your Potential SELF waiting for this very question to call it out and *actualize* it.

EXPRESSIONS HELP BECOMING

When I act on things outside my SELF, I change them somehow. Contact with me leaves them printed in my personal style with my own mark. I leave my marks even on things I destroy. This mark in itself is an extension of my being. It is as if I had written my name, my identity, on something. By this extension I have become more and it pleases me.

Show me an old school photograph. My eyes will race over the faces of old friends in search of my own face. Show me a photograph of the President of the United States shaking my hand and giving me a medal. Will I look at the President? Or the medal? Neither. My eyes will go to me on the picture. That likeness is my mark.

I can become still more if my mark, my name, is read and recognized by other persons. My mark lets me live in their minds and feelings.

The human being has a deep longing to be known and remembered. He fears ingratitude and rejection. He fears the erasure of being forgotten. If sides are chosen and his name is last to be called, it is still better than not to be chosen at all. Recognize him, remember his name, recall something he's done and you give him genuine pleasure.

This is what man wants most, to be and to be more. Satisfy that want and you satisfy man.

Let him "mark" you. Meet him. Face him. Accept his hand. Respond to him. Smile.

If he talks, listen. If he acts, watch. If he tries to please you, be pleased. If he makes a chair, sit on it. If he cooks, eat. If he writes, read; paints, look. If he grows flowers, enjoy them. Tell him you've walked on his bridge, that you've bought shoes from his shop, that you've visited a house he has built, that you admire his children. You assure him of his most profound success—he is not alone, he is in *you*.

A GOOD DATE IS A BECOMING

Experienced daters know that communication makes or breaks the date. Why is communication so important? Because *communication is the means of making ourselves known to another person as well as the means of making that person known to ourselves.* Without being truly known we can't be truly loved. Communication is our means of growing, of finding ourselves and finding love.

What do we communicate? Thoughts we have together and alike. Thoughts we have alone and different. Guesses. Worries. New words. Old words used a new way. Plans. What we used to do when we were kids. What we plan to do when we grow up. Feelings we want to share and feelings we don't know how to share. Agreements and differences. Changes of feeling.

INTRO

> The center of the date is the exchange between two people. The togetherness which is the purpose of a date is strengthened and enhanced by the communication.
>
> *Girl, 18, Milwaukee*

> We both went home a little happier to be alive. Part of that person went with you. That was a good date.
>
> *Girl, 17, St. Louis*

> If you can say you knew a little bit more about your date than you did previously, then nothing's been wasted. A good date is like getting to know you.
>
> *Girl, 18, New York*

> A good date is one where I believe the girl likes me more than she did before we went out. A *good* date is one where both people concerned have shown each other a good time maturely.
>
> *Boy, 18, New York*

> I would consider a date successful if both parties had a good time and neither party thinks less of the other (Perhaps even if both parties thought more of each other—even if it is just the sense of a "friend"—a good friend).
>
> *Boy, 17, Milwaukee*

> A successful date would be one when the boy and girl can speak more freely and can become better friends.
>
> *Girl, 18, Denver*

> . . . when the boy and girl both go home and are very happy and content with themselves and the other party. This type of date would leave you both with a feeling of respect for the other party.
>
> *Girl, 17, New York*

> Sometimes communication isn't necessary such as on the way home from a date. Driving home you are both tired and enjoying each other's company and usually listening to the radio. On a date it is different, especially a first date. You want to learn more about each other, therefore you exchange bits of information.
>
> *Girl, 18, Milwaukee*

A BAD DATE IS NOT-BECOMING

A bad date is simply one you didn't enjoy and will try to forget.

Girl, 17, New York

The person that you go out with has to be concerned about you as a person and really want to get to know you. You will never really get to know or be close to a boy who you can't just sit down and talk to.

Girl, 18, Milwaukee

Some girls can sit next to you all night in a movie house and if they don't wear perfume or chew gum you don't even know they are there.

Boy, 18, Denver

The two sit there all evening with nothing to say, each with his own thoughts.

Boy, 17, St. Louis

He picks you up; you go to a show and then come home without any conversation whatsoever. The evening is drab and boring and you end without knowing who you had the date with.

Girl, 17, St. Louis

Both parties go their merry way and don't have fun with each other, only one contributes to the evening and the other sits back and wants to be entertained.

Girl, 18, Milwaukee

The guy stays far away—he's not really there.

Girl, 17, New York

I used to go with a boy and the reason we broke up I guess was we couldn't talk things over well or seemingly get ideas across. We had good times, but it seemed as if two different people were doing the same thing by accident rather than plan.

Girl, 18, New York

It was as if the date really never existed. I remembered nothing except it was the worst date I've ever been on.

Girl, 18, St. Louis

The guy couldn't talk, he would sit there and mumble to himself instead.

Girl, 17, Denver

A bad date is when you feel uneasy with a person. You feel you must be talking constantly.

Girl, 18, New York

A bad date is when the person one went out with just won't pick up the conversation no matter what you try to talk about. It is a miserable time because one doesn't find out anything about the other party.

Boy, 18, Milwaukee

I went out with this guy on a fix-up. My girl friend said he was darling. All I knew was that his name was Gary. Well it turned out he wasn't darling, he was homely. That doesn't matter much but it started everything out wrong. I was in a bad mood because I was tired and sunburnt. He didn't talk we just sat there. We went bowling and didn't talk. When I came home all I knew was that his name was Gary.

Girl, 17, St. Louis

Self-Image

Three fat boys, Walter, Chuck and Marty board a bus. There's an empty seat between two people. Walter glances at the seat and decides he'll stand. He knows he won't fit.

Chuck is thinner than Walter. He could fit. But he sees two people looking—he imagines they are glaring at him. They're unfriendly, he decides. They don't want him to sit. He looks away. His face turns red. Marty, who's bigger than Chuck, sees the space and plops into it. He pulls a sandwich out of his school bag and starts eating, filling the space even more.

Walter, Chuck and Marty each have different body images and they behave accordingly. Walter has a proper body image. Marty knows he's big but doesn't care what other people think. Chuck has been made to feel his fatness is ugly. He suffers from "B.S.I."—Bad SELF-Image.

Every person has a *body image*. It is how he consciously and unconsciously sees his own body.

Every person has a self-image, too. The self-image is the person's understanding of his Actual and Potential SELF. It is the fullest description a person can give of himself at any given time. He would describe not only what he knows but how he feels about himself, whether he likes himself or not, whether he's self-assured or fearful. The self-image contains his body image but includes more. It adds psychology to physiology. It includes his thinking ability, his emotional capacity and temperament, his power of imagination, his ability to remember, his power of logic and reasoning, his ability to attract and persuade others, his likeability and popularity.

The body image tells a boy whether a hat is too big or a belt too small. He can identify himself in the Year Book through his body image. He can recognize his voice on tape because of it. It's what tells him whether to try out for football or play chess instead.

A girl has some idea of her talents, whether she's good at math or languages, whether she can drive a motor bike or not, whether she's cool or flustered in a crisis. She knows whether her decisions work out right most of the time or whether they often get her into trouble. Her image of her inner resources have a lot to do with whether she goes to college or not. She consults this same image while she reads want ads looking for a job. A girl's self-image may be so poor she can't picture herself either in college or finding a good job. She may conclude all she's fit for is immediate marriage.

The self-image has an extremely important effect on our lives, perhaps for the good, perhaps for the bad. A good self-image can free us to meet and win against towering challenges. A poor self-image can put us down without our even trying, paralyze us emotionally and intellectually, nail us to the ground when we have the talent to fly.

Self images have this power because they affect our two most vital arteries for *becoming*—communication and decision-making.

HOW B.S.I. FREEZES COMMUNICATION

I was asked to escort this girl to a private school (high class) dance. Lack of car and money made me feel uneasy. I was afraid of what to say.

Boy, 17, St. Louis

I was in a very bad mood because my hair didn't turn out right and I underdressed for the occasion. I was hard to get along with because I felt like such a grub.

Girl, 18, New York

You need communication to just meet the girl and to start talking to her and I feel this is my way of losing a date before I get started.

Boy, 18, Milwaukee

On my first formal date I was very self-conscious about the way I looked, talked, and acted. My date must have been the same way, because whenever we would start talking we would begin stuttering and stumbling over words and never finished what we were going to say. So we just sat like lumps on a log all night.

Girl, 17, Denver

I went out with this guy who was so concerned about doing the right thing, he wasn't his natural self. This spoiled the date because he was afraid he would say the wrong thing.

Girl, 17, St. Louis

If you are really comfortable with a boy, it really doesn't matter how much you talk, communication can be achieved in other ways—mentally.

Girl, 18, Milwaukee

If I'm with someone I hardly know, or if I feel I have to come up to the level of his "reputation" then I'm uncomfortable. I feel I have to talk, and every word must dazzle him. Of course, this rarely happens and so of course, I usually feel like I'm boring him to death, which makes me sound like more of a dim-wit when I continue to speak. This is where the lack of communication occurs. I have no confidence in myself unless the boy is more bashful or awkward than I am. It's probably because of this feeling, that I hate to be thrust upon a boy, matched or arranged as a blind date or something.

Girl, 17, Milwaukee

Dr. Morris Rosenberg, a social scientist, after examining the self-concepts of 5,024 high school juniors and seniors from ten high schools in New York State, separated the students into *egophobes* and *egophiles*.

The egophobe [ee-go-phobe rhymes with lobe] is a person with extremely low self-esteem. He's a self-hater.

LOW SELF-ESTIMATE: the egophobe rejects himself, is dissatisfied with himself, has self-contempt. He lacks respect for the self he observes. He feels he is inadequate, unwelcome, ugly, dumb, incompetent.

The egophile [ee-go-FILE] is a person with extremely high self-esteem. He's not blind to his defects but he likes himself.

HIGH SELF-ESTIMATE: the egophile respects himself, considers himself worthy. He does not necessarily consider himself better than others, but he definitely does not consider himself worse. He does not feel that he is the ultimate in perfection. On the contrary, he recognizes his limitations and expects to grow and improve.

There are few *pure* egophobes or egophiles. All people will have some of both characteristics. A boy may be modestly proud of his athletic ability but embarrassed about his dancing. A girl may want all her photographs taken at a three-quarter angle but refuse to pose for a profile.

But some persons have more of one characteristic than another. It is these we will describe as egophobes or egophiles.

Dr. Rosenberg discovered that leadership and a good self-image usually went together. The egophile had confidence in himself, so others had confidence in him too. On the other hand, egophobes, were more easily led, more easily influenced by others. Egophobes usually gave in to others and let others make decisions.

Like when I'm in school and somebody says "I don't like the clothes you are wearing." So I say that I am

going to stop wearing the clothes, and won't wear the two pieces together again.

From *Society and the Adolescent Self-Image*

Egophobes rarely or never engage in discussions of student government or general high school interest. They are passive in discussions of high school matters. They are least likely to be asked to state their opinion on school matters.

The self-image is contagious. Whether good or bad, others pick it up too. The egophobe makes others dislike him or lose confidence in him too.

> I like to date a boy who doesn't take himself too seriously and forgets about what the girl will think of him. I don't mean he should be impolite, but he shouldn't be just apologetic.
>
> *Girl, 18, Denver*

> A bad date is a person who constantly complains and shows little respect for himself or me.
>
> *Girl, 17, St. Louis*

> When a boy keeps asking the girl if she's having fun, he only makes you more on edge.
>
> *Girl, 17, St. Louis*

> I can't stand the guy who likes you so much he acts as though you are doing him a favor going out with a clown like him and keeps saying sorry for everything he does.
>
> *Girl, 18, New York*

THE EGOPHOBE DATE IS A LOSER

The egophobe boy is afraid to ask for a date because he's quite sure he'll be turned down. If he does try, he asks by saying, "You don't want to go out with me . . ." The egophobe believes that every bad breath ad, greasy hair ad, body odor ad, "Hey, skinny!" ad is directed at him. He stays home cringing in front of the television set.

The egophobe girl makes herself unattractive by her defeatist, pessimistic, sour, negative attitude. She is convinced that she's basic ugliness, basic dullness, basic unattractiveness, basic blah. This girl may actually be very attractive. She may have an excellent sense of humor, a good voice, a good figure. She may be talented or highly intelligent. She may be quite beautiful. But none of this matters.Nor does it matter that other people think she's beautiful. She can't accept herself. Her own image of her SELF is corrupted.

An inferior self-image doesn't make all individuals retreat into themselves. Some egophobes are aggressive and go about trying to prove they're really better than anybody thinks they are. Some criticize others to save themselves. Some clown around to distract attention from their real or imagined defects. Some boast and lie attempting to build up a favorable image through words instead of actions. An egophobe believes that when you praise someone else you are really attacking *him*.

> One of my dates seemed not to be the least bit interested in me. The whole evening was spent listening to him bragging about himself.
>
> *Girl, 17, St. Louis*

> My date happened to be the silent type, very rude and inconsiderate and impolite, and a drinker which made him boisterous. He acted wild and immature and tried to take advantage of me. It made a lousy evening.
>
> *Girl, 17, Milwaukee*

> Took a girl who talked about herself and hardly listened. Criticized too much and, in general, kept me out of the whole scene.
>
> *Boy, 18, St. Louis*

> Either he talks about himself all night or he's too shy to even talk to me. You can't win with that one.
>
> *Girl, 18, Denver*

SEVEN BURDENS OF THE EGOPHOBE

1. The egophobe is touchy, supersensitive, easily and deeply hurt by criticism, blame or scolding. He can't stand being laughed at. He is quick to shift blame from himself to somebody else or to circumstances outside of his control.
2. The egophobe finds it hard to meet new people, to open conversations and keep them going.
3. The egophobe is often deeply pessimistic. He doesn't try anything because he's convinced he's going to fail.
4. The egophobe assumes others think poorly of him and don't like him very much. He feels generally passed by and passed over, misunderstood.
5. The egophobe doesn't trust people. He has little faith in human nature.
6. The egophobe tends to hide behind a front. He adopts a special pose to meet the public. He plays roles in conversation rather than speaking from himself.
7. The egophobe is lonely and feels relatively isolated. He doesn't know how to give. He doesn't share himself with others freely, fully and spontaneously. He doesn't feel anyone wants to share any part of him.

Where Do Self-Images Come From?

Do I stay in my seat at dances? Cram against the wall? Freeze or pretend to collapse or make a big joke of it if a girl looks as if she wants me to dance with her?

Why? I can't dance.

But what is there to dancing? The ability to move one foot out before the other, to slide forward and backward, to bend, dip, peck, kick, jump, wriggle, flap the arms. There's nothing here that I don't do out on the field running after a ball.

It's because I have no rhythm.

But that's foolish too. Do I jam a revolving door? Do I have trouble coordinating with an escalator? If my timing is good enough for a revolving door, it ought to be good enough for dancing.

I'm shy. I don't have enough confidence in how I look. I'm afraid I'll make a fool of myself. I'm afraid people will laugh. I'm afraid they won't approve of me, won't accept me.

All right. I don't like how I imagine I look when I'm dancing. I have B.S.I. (Bad Self-Image).

Where did I get this poor image of myself. Who let it in? Very likely I got it back down in the wee years of my life.

We never see ourselves in a vacuum, isolated from other persons. Always with all of us, just to one side, nearby, in the background or foreground, we sense the presence of others, a shadowy audience, watching what we do, writing down what we say. It exerts a powerful influence over what we do, this audience. We fear it or love it. We want its applause. We are always showing off before it, hoping to impress it. Or we resent its presence and do things to spite it. Or it's a never-satisfied, ever-complaining critic and we so dread it we can't move. Usually the presence is nameless and faceless, but sometimes we recognize a face or voice. It is mother or father. A teacher. A sister or brother.

The infant begins to experience his body while still in his mother. After birth, as the baby grows, he starts using his hands to explore himself. He sees that he and his mother are not one body. He discriminates between himself and others, always using his own body as his standard reference object. As he grows older and becomes aware of his thoughts and feelings, as he talks with others in his family, his self-image grows the same way. He learns his limits by meeting resistances. He learns his power by extending himself, by probing. He is forever testing his value by testing how much parents care for

HOW DO YOU COPE WITH NON-LISTENERS?

him. How much can he get away with? How far will they let him go?

The child's image grows for good or bad according to how he gets along with the most important people around him, his mother and father, his brothers and sisters. The attitudes of his parents decide for him whether a body part is clean or dirty, attractive and pleasing or ugly and repulsive.

The child whose body structure satisfies his family is usually neither over-sensitive or under-sensitive about his body. He accepts his body for what it is and lets it go at that.

Some parents over emphasize muscle prowess and strength in boys or they may give unrealistic, exaggerated attention to prettiness in a girl. There are parts of the body that some parents play down. They avoid talking about these parts and keep them covered and so may give the child the feeling that these are parts to be ashamed of. The feeling of shame may last through life.

Now about my dancing. Maybe I walked funny and it amused my parents. They thought it was cute. They called attention to it: "Look at that walk!" they said. "Will you *look* at that walk." I thought they were laughing *at* me.

The mother's own attitudes and feelings about her child's body will have the most to do with the self-image he forms. Many mothers overrate the importance of physical beauty. If a little girl has some defect or disfigurement, the mother may feel humiliated, depressed or guilty, as if she had done something wrong during pregnancy. These feelings will make it hard for her to completely accept her child and the child will sense it.

A mother who is unhappy about her own looks and unsure about her own likeable qualities may feel very disappointed if her infant daughter isn't a beauty contest winner. If the little girl has big ears or thin hair or a weak chin, the mother may unconsciously reject her. There are any number of ways this rejection can take form. She may spend hours fluffing up her hair or she may comb the hair down over her ears and teach her to keep her ears hidden. She may unconsciously give more love and time to another of her children. There are also any number of ways a child may react to the rejection. She may become excessively shy, for one. Shyness here is nothing more than imitated shame. Since the mother is not proud of showing her off, neither is the child proud of showing herself off.

> I was afraid to ask questions. When it came time to know about "the facts of life" my mom took me aside and made it very secretive and mysterious. She was embarrassed and uneasy. When it came time for me to wear a bra and all, my brother and sister, were allowed to

tease me all they wanted to. My body embarrassed me when it was changing.

Girl, 18, Milwaukee

A person may have a fair appraisal of his body and personality but may still be excessively shy. It isn't his *self*-image which is defective but his image of *acceptability*. He thinks: "I'm all right but they won't like me." He is ever ready to believe he isn't wanted. He expects others to put him down. He may boast for this reason. He *over*-rates himself because he expects to be *under*-rated. In conversation, he talks loudly and rapidly in fear of being cut off.

How did he get this way? It's impossible and silly to expect one answer to explain all braggarts, but a good many individuals brag because nobody at home ever cared about what they said.

A person with a good self-image feels no need to boast or brag or to pull himself down. He knows his abilities and he states them with assurance and confidence.

The braggart doesn't really expect anyone to believe him, hope so though he may. He is delighted when someone lets him talk himself out or accepts what he says without question. With such a person, he gradually boasts less and very gradually talks less and less. He begins to be interested in the other person.

How to Beat B.S.I.

Nobody needs to be immobilized by B.S.I. What can a person do to improve his self-image? Six things:

1. He can *pray*. He can seek and meet God. The worst self-image in the world must burn away in the image of God. None of the fears, weaknesses and failures common in human communication will survive communion with God. No one can keep thinking he's a creep and say he understands God's intelligence and God's love. The individual who understands what it means that God loves him *has to walk tall.*

2. He can develop a trust in people and love them as God loves them. He is a brother or sister, a son or daughter in the family of man. If he despises and mistrusts the family, he despises himself too for he is a member of that family. He respects himself by respecting others.

3. He can practice *dialog*—relate honestly and openly to people. He should try to talk to people, not his exaggerated image of them.

4. He can practice good decision-making. He should especially try not to *under*-promise or *over*-promise himself but rather commit himself to doing only what's in his power.

5. He can *compensate*. He can learn to make strength out of weakness. *Think positive*. If a boy is convinced he's weak in math, rather than dwell on the weakness he should develop language. If a girl feels she is ugly, let her at least be clean. There is no reason for anyone to succumb to ugliness. Mme. Helena Rubenstein's motto used to be: *"There are no ugly women—only lazy ones."*

6. He can try to *understand the origins* of B.S.I. Without excessive self-examination or without spending self-pity on an unhappy childhood, an individual can learn that misgivings he has about his body and appearance have little or nothing to do with the body he now has. Doubts and fears of his talents may not at all come from present abilities but from some dim earlier past.

Quiz

PART A

1. What is *rejection*?
2. What is *anxiety*?
3. How does anxiety differ from fear?
4. What is the human infant's first psychological need?
5. Describe four ways a child can react to rejection.

PART B

1. What is a human *drive*?
2. What do we mean by a human person's *central* drive?
3. What does "being" mean?
4. What does "becoming" mean?
5. List ten ways a person *becomes* more.

PART C

1. What is the SELF?
2. What do we mean by the *Actual* Self?

3. What do we mean by the *Potential* Self? What is the difference between Actual and Potential?
4. What is a *body* image?
5. How is a self-image different from the body-image?
6. What is an egophobe?
7. What is an egophile?
8. Can a person be purely an egophobe or an egophile? Explain.
9. What do the letters BSGF stand for? What does BSGF mean?
10. Why is adolescence called a "second birth"?
11. Name seven instances in early childhood which affect self-esteem for good or bad.
12. What do the letters B.S.I. stand for?
13. Name five ways to beat B.S.I.

CHAPTER THREE: DECISION-MAKING

Decision-Making: The Central Means of Personal Becoming

"If I choose, I am; if I do not choose, I am not."
Karl Jaspers

Decisions and the way we make them affect our health, our studies, our social life, our play. They make *personality*. They give us *style*. How we make them gradually wears on us; they color the way we stand and walk, scratch lines on our faces. They take their little share of food and energy from our bodies. They etch our personalities and mark each with our own distinct stamp, making each of us special, unique, different from anyone else. This stamp is *character*.

The Italian film actress, Anna Magnani, objected when a photographer touched up one of her portraits by "erasing" wrinkles. "It took me years to get those lines. Put them back. You've stolen my face."

Edwin M. Stanton, Lincoln's Secretary of War, remarked: "A man of fifty is responsible for his face."

Every day each of us makes an uncountable number of choices. Hundreds are so trifling that we aren't aware of making them and we'd laugh if anyone glorified them by calling them decisions. Should I get out of bed now or take five more minutes? What should I wear to school? Should I have my ears pierced? Should I do homework before dinner or after, or before school in the morning? Should I read or watch TV? What should I read? What should I see? Should I cut my hair?

Decisions are personal when they affect one person alone, myself. *Social* decisions, since they affect other persons, are harder to make. A girl's decision to accept a date will affect not only her but the boy who asks her. The boy's decision is still more complex. His decision was to ask her to decide. Marriage is a change from a life of personal decisions to a life of decisions which grow progressively harder and more complex as the marriage grows and as children are added to the family.

Taste is esthetic decision-making. Having bad or good taste really amounts to making bad or good decisions. A writer, musician, artist, dancer, a fashion designer improves in taste by repeated good decisions.

Sin is a bad moral decision-making. The ancient Greeks called sin "missing the target" comparing it with an arrow shot wide of its mark.

We have said that man's most central urge is *to be* and *to become more.*

How does he exercise this drive. What can he do about *becoming more*? Some individuals believe they *become more automatically* by *getting* more, by *adding* things to themselves, by collecting "status symbols." They pick their clothes for status. They pick friends for status. They marry for status. They drive racy cars or expensive looking cars. They drop names around like confetti.

It's possible to *become more* through these means but in themselves books, cars, money and property do not bestow *becoming.* An unimaginative amateur photographer with a bag full of professional gadgets will still take dull pictures. A dull writer who owns Merriam-Webster's Third International unabridged dictionary remains a dull writer. A bore in a sports car is still a bore.

Man becomes more man by the decisions he makes. Why is that? Because every decision involves at least one judgment; most pack a cluster of judgments. The judgment is the nucleus of the decision. It is this nucleus which distinguishes the human animal from other animals.

The judgment is so characteristic of man it is called the *human act.*

Epictetus told an athlete in ancient Greece: "If you're proud of kicking a ball far, know that you're proud of the act of an ass."

The philosopher was not ridiculing athletes. He was merely putting sports in their place. The human being's proudest act is not something a horse or dog or fish can do (often better). It isn't an act of speed or strength. It is *reasoning, judging, deciding* and *communicating.* Of course, man must reason and decide and communicate in competitive sports. The point is that only in judging and deciding does man act like a man.

The judgment is a simpler act than the act of deciding. In judging, we connect or disconnect two or more ideas or we associate an idea with an object outside the mind.

But decision-making is more than a mental operation. It involves the full person, *mind and will*, body and feelings too.

A boy *judges*: "This is an interesting girl." He *decides*: "This is an interesting girl. I want her and I'm going to get her. I'm going up to her right now to ask for a date."

The judgment *states*: "This is a Mercedes-Benz SL 190. This is *not* a Mercedes-Benz SL 190."

The decision *chooses*: "This is a Mercedes SL 190 but I don't want it. I want a Jaguar."

As the eye sees and the ear hears, the human mind must make judgments. As the body grows by eating, rest and exercise, the mind grows by judging. Every judgment adds something to us for in knowing we *become what we know.*

Judging makes us *more* but decision-making makes us still more. It calls for the full person to act. It engages the mind, the will and the emotions.

Decision-making also helps us become more through knowing. The decisions we make teach us *who we ourselves are*. We learn from them what we can do and can't do, what to attempt and what not. When we know ourselves, we can promise to act in the future. We can make contracts and commitments.

Just as we learn ourselves through decision-making, we also *learn to know others*. We recognize the phony and the boaster. We spot the timid or fearful, the confused, the unsure.

The human being expands and extends his name and his being not only by knowing more but in what he does to the world around him—his family, friends and school, his city and country.

The mental act of judging affects only the judgment-maker. The decision-maker affects himself but can also directly affect others. Judging is a personal and internal act. Deciding is an internal act externalized. The judgment is completed inside the judgment-maker. The decision strikes out of the person and changes the real world outside of him.

He extends himself into that world by his decisions. He extends himself by building houses and roads, teaching, fathering and mothering children, bowling, demonstrating, paying taxes and contributing to the support of his church.

Since the will operates in decision-making, the act of deciding is a freer, more versatile act than that of judging. Decision-making is an act commanding further action.

The judgment 'waits' for sensed information before it acts, but a person can decide to push out in search of new and different information. He can also decide to reject a judgment even if it is a good one. He can go against it, deny it, contradict it. He can decide to rationalize, that is, seek to construct judgments to support what he's already done.

Little Decisions Are Important

All decisions are attended by smaller decisions which serve as *pilots* or *satellites*. Someone blocks your way in the corridor, for example. You can pass to left or right, collide, stop or retreat; you can smile, grumble or apologize and while you're doing that you can shake a fist or a finger or you can wave or push. What you decide is going to be affected by where you're going and what your mood happens to be. If you're late for an important class, you'll be impatient. If you're running to report a fire in your clothes locker, you may become savage and bowl everybody out of your way.

Small decisions, besides serving their own objectives are important in other ways. They keep our decision-making mechanism in condition. How we make small decisions will have very much to do with how we make big ones. It's by small decisions repeated with attention that we create the habit of mind which lets us act speedily and smoothly in a crisis. It's for this reason we have rehearsals before opening a play, practice games before the real game, boot camp and basic training before battle—to exercise decision-making. It's for this reason, too, that we are novices before taking religious vows and have engagement periods before marriage. Many of the decisions of marriage get tested during the dating and engagement period.

Edith McFaum was trailing her husband's five ton horse van down a Connecticut mountainside when she saw oily, blue smoke streaming from under the rear of the truck. She raced her convertible alongside and looked over toward her husband. His face was white. He was holding the steering wheel with his left hand and, with his right, yanking the handbrake.

"My brakes are gone! Get out of the way" he yelled.

Edith McFaum dropped back and followed her husband as the van picked up momentum.

It was Sunday, July 24, 1955. A church at the foot of the mountain was ringing its bells. Hundreds of people would be in front of the church, saying hello or goodbye or standing to talk with good friends. What would happen when the van came hurtling down on them? It didn't even have a horn.

John McFaum could have steered off the road into a field but he wasn't alone in the van. Two young grooms were back in the van with the horses. If the van rolled over, the horses would crush them.

Edith McFaum cut in front of the van, her own horn screaming alarm down the steep winding slope.

Half a mile ahead two cars waited for an intersection light to change. The first car had its directional light blinking. It was going to make a left turn directly into the van's path!

Mrs. McFaum pushed her convertible down to the intersection and tore up a grassy embankment where she waved and begged the turning driver not to move. He looked at her in amazement, and began to make a distracted looping turn, his eyes on her instead of the onrushing van.

He was barely past the center of the road when the van hit the intersection. John McFaum at 60 miles an hour swung cleanly around the tail of the car turning left, then back into his own lane.

His wife backed off the embankment and raced after him. At the next turn in the road, she cut across the grass and gravel and once again nosed to the front.

Moments later a sign announced: "Winsted, Population 10,000."

Winsted's narrow main street was made still narrower by cars parked along the curb. A woman backing out from the curb stopped suddenly and killed her motor.

Midway through town they screeched around a corner and there was the church. A policeman stood on the sidewalk, ready to direct traffic. A newsboy stood next to a stack of Sunday papers. But the church doors were still closed. Not a soul was in the street.

Just past the church, the street split around a large grassy circle protected by a ring of iron pipe and chain. Beyond the circle, was a narrow bridge. Four or five cars were approaching the bridge from its far side. Mrs. McFaum raced to stop them, but too late, the first car was already on the bridge before she got there.

When John McFaum came to the circle he thought for a moment he would crash into the chain and post —they would stop him. He changed his mind when he saw there were no cars parked in the circle. He steered the van against the curbing. The air was filled with a terrible spitting, squealing sound. Blue smoke rising from his wheels, he stopped the van a bare twenty feet from the bridge. His face was the color of ashes and he couldn't talk. His fingers had to be pried loose from the steering wheel, and it took more than ten minutes to unlock his fingers from the brake handle. There was blood on the handle, he had gripped it so hard. John McFaum had driven the entire way one-handed.

Heroic courage like Edith McFaum's doesn't just happen suddenly. Great acts of love and important decisions are built gradually on a foundation of small acts of love and petty decisions. As Edith McFaum herself explained: "It was those fifteen years of getting up in the dark, cold winter mornings at five o'clock to make John's breakfast that got me ready for this."

Guidelines for Decision-Making

Decision-making is our most important means for personal growth and advancement. The guidelines below are steps to becoming a fuller person.

SIX POINTS TO REMEMBER *BEFORE*

1. Face it. Don't run away from making a decision.
2. Avoid making an important decision when you're fatigued or under the influence of an extreme mood—for example, deep gloom or wild excitement.
3. Assemble and write down the main facts.
4. Separate your most important purpose or goal from minor, satellite purposes.
5. Collect and grade your advice.
6. Be sure you have it in your power to carry out what you decide.

THREE POINTS *AFTER*

1. Trust yourself. Don't fret over your decision once it's been made.
2. Be able to change your mind when a change is reasonable.
3. A decision is only half-born if it stays in your mind. Tie every decision to some practical action. Decide, then do.

Before Deciding

STEP 1. DON'T RUN AWAY FROM MAKING A DECISION.

"Never put off till tomorrow what you can put off till next week."

Frank Sullivan

We may avoid decisions by postponing them, by flipping coins, by passing them on for someone else to make. But the person who avoids a decision does, in fact, make a decision—he *decides not to decide.*

A Brooklyn girl, afraid to marry, disappeared on the morning of her wedding. Her parents and bridegroom thought she had been kidnapped. Police searched the parks and beaches expecting to find her body.

More than a month later she walked into her house, tired, thin and nervous. She had been riding buses for a month. She had traveled from New York City to Los Angeles and back, the entire route on local and short-haul buses. All this because she couldn't make up her mind to marry.

STEP 2. AVOID MAKING A DECISION WHEN YOU'RE FATIGUED, EXTREMELY HAPPY OR EXTREMELY UNHAPPY.

Fear, laziness, jealousy, anger, hate, melancholy, lust—any of these feelings can affect a decision. Alcohol can block a decision or distort it.

A Jesuit novice master once said: "Most novices decide to leave the Order in winter at five in the morning—the hour they have to leave their warm beds and face the cold day. By breakfast they decide to stick it out a little longer. By eleven they've forgotten their troubles and are writing friends to join."

A tea taster may be able to identify as many as fifteen hundred different teas. From taste he can tell the district where the tea is grown, the time of year it was plucked, sometimes even the owner of the field it comes from. But his taste buds are effective for only a few hours. He never judges a major lot of tea when his mouth is tired.

According to studies by the National Safety Council, the majority of fatigue accidents on highways occur after 250 miles of driving. Driver fatigue studies in Chicago showed that the average driver's efficiency improves during his first three or four hours on the road; after six hours he will usually fail reaction and performance tests even though he may insist he doesn't feel tired.

A man or woman is not the same at 5 A.M. as at 11 A.M., or 3 P.M., or midnight. Body temperature falls at night, dropping to its lowest point between 2 A.M. and 5 A.M. When body temperature drops, a person usually feels sleepy and dull. He's bored. He's unhappy. If he's driving a car or truck at this time, he had better pull off the road for coffee or a nap. Auto accident insurance companies report dusk is a danger period.

Professional decision-makers all give the same advice: don't try to make a big decision when you're high or when you're extremely low. Get rest, then decide.

STEP 3. ASSEMBLE THE MAIN FACTS INVOLVED IN YOUR DECISION.

Writing facts down lets you see them all at once. You can arrange them according to their relative importance. If you try to juggle all the pertinent facts to make an important, complicated decision, you tire your brain. The final decision will be test enough without burdening yourself meanwhile with memory tests.

Fact sources are determined by the kind of decision to be made. If you're thinking of going to college, you will have hundreds of facts to collect and compare before you decide. You will get some school catalogues and study their various entrance requirements. Do your grades make you eligible? Do you qualify for any scholarships?

It's sometimes a mistake to try to collect *all possible* facts related to a decision. You may be putting off the actual decision. The researcher who can't stop researching isn't necessarily in love with truth. He may merely be afraid to write a report.

STEP 4. DEFINE YOUR GOALS.

Often, while making a decision, you will find you're confused because you're trying to satisfy several conflicting goals at the same time. Once you distinguish these various objectives, the most important is likely to stand clear of the others and your decision will be easier.

A policeman won $16,000 in a television quiz contest. He decided to drop out and gave this explanation: "I weighed the excitement and vanity of winning against the security of my wife and five children." The policeman's decision is a clear case of choosing a major goal in the face of a tempting conflict-goal. A woman in the same contest quit at $32,000 because her doctor advised her that going to the limit would put too severe a strain on her health. Life itself was her major goal, the fun of gambling and fame was her conflict-goal.

It's extremely important to separate reason and fact from feelings. Once separated, try to find out the origin of feelings which conflict with the stated major goal. A conflicting feeling may be evidence of a conflicting goal. *The decision should weigh goal against goal, not goal against feeling.*

STEP 5. COLLECT AND GRADE YOUR ADVICE.

When President John F. Kennedy decided to make Glen Ora near Middleburg, Va., his country residence, he embarrassed certain local Catholic leaders, for Middleburg did not have a Catholic church. Mass was said in the town's community center by Father Albert F. Pereira, who came from a nearby town.

Father Pereira's bishop asked him whether he would think of saying Mass privately at Glen Ora for the Kennedy family.

"My immediate opinion goes against the private Mass," he said promptly "but I'll think about it and let you know in a day or so."

A few days later the priest put his question to some Protestant ministers at a community meeting.

"How would Protestants react if I said Mass privately for the President at Glen Ora? Since the United States is predominantly Protestant," the priest reasoned, "I thought it would be wise to get your thoughts on the subject."

There were some moments of raised-eyebrow silence as the ministers warily studied the priest and glanced around measuring each other. "Some of you didn't vote for Mr. Kennedy," Father Pereira urged, "but I'm sure none of you would want to see the President embarrassed by a religious issue that could possibly be avoided."

No one spoke up at the meeting, but privately later, the ministers one by one came to Father Pereira and explained their ways of thinking. They thought Sunday worship was a public act in which the community together as a unit praised God. A privately attended Mass, they predicted, would strike most Protestants as undemocratic. Some would accuse him of being ashamed of his religion. Some would wonder whether the Vatican was giving Kennedy secret orders.

Before Father Pereira left, he had the frank reaction he wanted. "Don't do it," the Protestant ministers warned.

President Kennedy attended Mass in the community center.

STEP 6. BE SURE WHAT YOU DECIDE IS IN YOUR POWER TO CARRY OUT.

A good decision requires a proper self-estimate. I should know my voice before I attempt to entertain others by singing. I should know my hunger before I serve myself food. I can hurt myself if I try to lift a weight heavier than my strength can manage. I should know how much time I have for reading before checking books out of the pay-library. I will not apply to become an airline hostess if I'm four feet ten inches tall and weigh one hundred and sixty-five pounds.

A good decision requires SELF knowledge. How do I know myself? By watching myself, listening to myself, noting success and failures.

I learn to know myself also by asking others what they know of me, by watching reactions to my actions. I can ask and accept advice openly only if I'm ready to discover that I am weak and can make mistakes. Feeling or intuition alone is no sure way of evaluating one's personal capacity. Edith McFaum's decision to drive down the mountainside ahead of her husband was not based solely on love and intuition. She was an able, experienced driver. Her confidence matched her desperation.

By contrast, here is a statement about a man who *over*-promises himself. The statement comes from a mother asking for legal separation on grounds of non-support:

> He always told me he loved me so much he'd *die for me.* I got sick of hearing it. I didn't want his death. All I wanted was for him to get a job. Go to work every day and pay our bills. He wanted to die for me, he said. He couldn't see dying was the cheap way out.
>
> Milwaukee, Family Court Proceedings.

After Deciding

Once we've made a decision, we can do three wrong things to it: worry it to death with doubts, second-guesses and fears; go ahead with it even though we begin to see it's wrong; or, keep it in mind as a good idea and never let it see daylight as an action.

1. MAKE YOUR DECISION AND LEAVE IT.

In any game where a ball must go over the plate or through a basket or between upright bars or into a hole to score points, our control over its destiny endures in the brief moment of contact with the ball before we send it on its way. The contact may be cold, as with the steel head of a golf club or violent as with the toe of a football shoe, but the contact is critical. Is the grass wet? Are strong cross currents in the air? These factors in our mind have to be imprinted on the ball while we have it. We can't ride along with the ball and no amount of "body English" will reach it once its gone.

The baseball umpire is a master at dismissing a decision from his mind once it's been made. A National League umpire described his decisions this way: "I make up my mind fast and call the play hard as I see it. If I hesitate, the crowds will think I'm not sure. They'll start throwing bottles and the players and managers will start roaring down on me. There's no second chance. Apologies don't work. You set your jaw, stiffen your neck and close your ears to everything. If anybody disagrees too loud, throw him out of the game."

2. CHANGE YOUR MIND WHEN REASONABLE.

The umpire's advice is bad advice if applied to all decision-making. We should be flexible enough to change our minds when a change is called for.

Our court system generally protects us from unfair court decisions. We have the right to appeal to a higher court. We ought to give ourselves the same protection against our own errors.

3. DECIDE, THEN DO.

On paper, a decision is a noun with its verb. I go. I do. I act. I promise. In life, a decision must be acted out to be complete. It is still a matter of I go and I do but now the I is not on paper, it is the *person* who must go and do. Many of our good decisions flop for no other reason than that we don't feel like carrying them out. Anticipate these drag feelings. Add positive feelings to support your decision.

For some people decision-making is a form of daydreaming. The daydreamer makes heroic decisions to work harder, give up smoking, go to church more often, study, but he doesn't dream through to the steps he's going to take for achievement. A man decides to quit smoking but buys a carton of cigarettes "just to have around the house." A housewife decides she's watching too much television. She's going to give her set a rest for awhile but she studies *TV Guide* to make sure she won't miss anything good.

How To Test Your Advisor

Proverbs are compressed experiences. A Spanish proverb says *"A proverb is a little gospel."* What do the proverbs say of advice?

Arabian: Never give advice in a crowd.
French: The well-fed give better counsel than the hungry.
English: Advice is least heeded when most needed.

What do writers of the past say?

Lord Chesterfield: Advice is seldom welcome; and those who want it the most always like it the least.
Nicholas Boileau: Sometimes a fool makes a good suggestion.
Leigh Hunt: Advice is not disliked because it is advice; but because so few people know how to give it.

The wisdom of the past offers this advice about advice: *Be open about needing it. Be sincere in asking for it. Listen. Consider the advice objectively. Overlook any show of superiority or arrogance in the giver. Look beyond appearances. Be able to trust.*

In seeking advice, we should be sure it is advice we want. Some people enjoy asking advice because then they can be the center and object of discussion and attention. For some, asking advice is a way of opening a conversation. Others don't sincerely want advice but want permission or approval of something they already plan to do. There are persons who ask advice merely for the pleasure of rejecting it.

Suppose we need help. We can't accept just anybody's advice. Neither should we look for someone who is sure to agree with us. How do we know whom to trust? What are the tests of a good advisor?

The person of whom we ask advice or an opinion is subject to the same tests for making a decision as we are. The counselor or advisor is making a decision too. Does he answer off the cuff? Does he hear you out or can you tell he's not listening? Does he seem interested in having all the facts before he answers? Is he tired or worried about something else? Does he seem very blue or is he giddily excited about something? His mood may affect his answer. Better wait and see him when he's closer to normal.

Is he ready to change his mind if he gets additional information? Is he willing to consult other authorities for their opinion or does he feel his opinion is utterly right and cannot be affected by anyone else's ideas?

Can a priest or a nun, or any single person give advice on marriage? Some say "no." (Some go further and say no one can give advice on marriage until he has been married and divorced.)

A monk who lives in silence in a secluded cloistered community, who never sees or talks with families, can still be an excellent marriage counselor or strike arbitrator or peace negotiator. Marriage is a relationship. In a very rare way, silence teaches the subtleties of inter-human communication for it teaches man to listen to the inner man.

1. The advisor only advises. He does not decide because he does not act.

2. The advisor should be willing to advise even though telling the truth may be awkward or embarrassing to him or others. Don't ask a coward.

3. The advisor should not let his counsel be subject to his personal mood.

4. The advisor should know pertinent facts or have an intelligent means of getting them.

5. The advisor should know your needs, goals and the best means to achieve them. His personal interests ought not conflict. He shouldn't have an ax to grind.

6. The advisor should be ready to check his information and his advice with other informed persons.

7. The advisor should be competent. He should not attempt to handle problems over his head. His ability should be matched with the advice he is expected to give. Honesty is a pertinent quality in a banker, not necessarily in a plane pilot. The pertinent ability we look for in a plane pilot is the ability to take a plane up and bring it down in one piece. St. Teresa said she would want an intelligent man as her spiritual advisor, rather than a man whose only quality was goodness.

The First Decisions In Life: Why They Mean So Much

In the early months of life, each new set of muscles means exhilarating fresh powers for the growing child. He can touch, clasp, kick and push. He can pick up and hold on, grab and let go. At first he blunders into these basic actions, but he learns to repeat them and is rewarded by the pleasures of achievement.

Once a child is *able* to do an action, he is *able to decide whether to do it or not.* He can touch *or* not touch. He can hold *or* let go. These primitive decisions are extremely important. If the child doesn't practice them, he will suffer for it all through life, just as the child who never learns the alphabet will never read or write. He learns better and faster if he's coaxed and encouraged. If he's discouraged, he may not learn at all.

If his parents are confident in him, the infant will learn confidence in himself. If they're over-protective or impatient, they'll interfere with his clumsy trial and error experiences and so take away his chance to learn. He will *feel* their unwillingness to trust him and may consequently distrust himself.

If the child doesn't trust himself and is afraid of failure, he may give up and not even try. He may react in a dozen other ways. He may repeat things he already does well because he's sure of his reward. He is starting to talk, for example. Instead of trying new "words," he may repeat his sounds almost hypnotically. He may stammer. He may throw screaming, choking fits of rage and frustration.

Whatever the form or manner of a child's reaction, if his parents repeat their behavior, he will repeat his and soon his reactions become habit. This habit or a variation of it will have to be reckoned with later in his home, in school and ultimately in his marriage and work.

Children, like puppies, have to be housebroken. So, usually between the ages of two and four, they are trained to use the toilet. When they discover they have some control over the process of eliminating, they derive pleasure from this achievement just as they do from others. They may also discover that father and mother seem especially eager for them to eliminate in a certain place and, if they do so, they can count on praise. If they don't use the right place or if they don't eliminate at all, they're also sure of getting special attention. The child thinks: "Should I do what *they* want or what *I* want? If I do what I want, they'll stay and beg me and make a fuss over me." Or he thinks: "If I don't do what they want, they'll leave me." Or he thinks: "Better do it their way. Keep them happy."

Doctors rate the toilet training period high in importance because elimination is a vital human function repeated for all of life. Adults have mixed feelings about this function and tend to communicate their attitude toward it to the child. Since the function is repeated, child and parent are brought together daily and regularly. Other functions bring child and parent together and they provide tests and exercises in communication too: the family meal time, shopping, exchanging holiday presents.

Some children try to please their parents by being scrupulously exact in obeying them. Later in life they will try to make scrupulously exact decisions. The exactness of their mental attitude is visible in the way they dress and keep house or in their work or speech. Their taste is scrupulously proper and so is their morality.

Such persons may be unable to reverse a decision once it's been made. They launch decisions like torpedoes which can't be recalled. Such persons may also fret a decision to rags after they've made it. They may avoid making decisions because they're afraid of failing. Of course, they're bound to fail because their standards are superhuman and impossible.

We make the decision to mind or not mind our parents so many times that the way we make it becomes almost the foundation decision of our personality. It marks the pattern of how we act toward authority and the rights of others, whether we conform or rebel.

The five most serious decisions which confront a person before the age of twenty all relate to some form of authority or to somebody's rights.

1. Should I finish school or drop out and get a job? If I decide to stay, what course will I take? Should I pick out a snap "Mickey Mouse" course or a tough, demanding one?
2. Should I stay home with my parents or leave?
3. Should I live by a religious code or not—keep the Commandments or not?
4. Should I think about marrying now? Should I go steady?
5. Should I obey driving regulations and laws restricting drinking—how much respect should I give to community authority?

Psychologists believe that how a teenager handles these decisions can be traced back to the person's early relationship with his parents.

Identification

In his early years, the child starts absorbing many of the standards, values and morals of his parents. Psychologists call this process *identification.* They say the child is *identifying* with his parents. The values and standards help form the person's *super-ego.* They are important in decision-making because they act as the background reasons for making judgments. They produce the purposes and the goals for work and play and living itself. They are important to communication because they help form the *I*, the *self* to be communicated. *Identification* differs from *imitation* in that identification is an unconscious process, imitation a conscious one.

DEFENSE MECHANISM: A psychic process, by which a person unconsciously tries to escape anxiety and avoid emotional conflicts.

IDENTIFICATION: A defense mechanism in which an individual endeavors to pattern himself after another. Identification plays a major role in the development of one's personality and specifically of one's super-ego, including the conscience.

The child accepts his parents' decisions as the best possible. If his mother files her teeth to needle points and pierces her nose with a bone, he accepts this esthetic decision as his ideal of what is feminine beauty. What his mother cooks is better than what anyone else can cook. His father is stronger than anybody else's father. His father can drive better and smoke better than anybody. If his parents cheat the landlord and steal from the supermarket, cheating and stealing will be his standard for moral action.

So the child begins to take on his parents' ideas. For better or worse, whether they are screaming at each other or buying each other surprise presents, reading or telephoning or watching television, his parents are the models he will imitate. Consciously (by imitation) and unconsciously (by identification), these values and judgments will referee his actions during his growing years, telling him when to act or not act, go ahead or stop. He may disobey these orders from his superego but he will pay for his disobedience with guilt.

EGO: This term should not be confused with "selfishness" or "self love." The ego represents the sum of certain mental mechanisms, such as memory and perception and specific defense mechanisms.

SUPEREGO: In psychoanalytic theory, the superego is that part of a personality we usually describe as the person's *conscience.*

The Image of Decision-Making

If we could watch ourselves televised in the act of solving a problem or meeting a crisis, we would see curious patterns of behaviour. We bite our nails. Our scalp itches. Our forearm feels suddenly tired. We stroke our upper lip or pinch our bottom lip. We stand numb, unable to do anything at all. We hesitate, look around for help and assurance. We tell everybody what we intend to do, then do nothing. We're furious and indignant when somebody tries to give us advice.

REGRESSION: In psychiatry, a partial or symbolic relapse into childishness on encountering difficulty. The regression may be shown by the position of the body taken during normal sleep. Regression appears in severe physical illness and in many psychiatric disorders.

If we track these attitudes and mannerisms to their beginnings their trail will lead us through freshman year, back down through grade school to our earliest pre-school years. For good or bad, this is the time we build the image of our decision-making.

Jim J. was always better at playing with toys than Jason J. He could unwrap a present faster than Jason and he could color with crayons better than Jason, too. When Jason couldn't get his tricycle to go because the front wheel was turned the wrong way, it was Jim who solved the problem by straightening the wheel. Jim, of course, was older and more physically developed than Jason. Jason was only five years old. His father, Jim, was thirty-one.

Jim never really showed Jason how to do anything for himself. He used to grab the sparkler gun out of Jason's hands to show him how to use it, but then he wouldn't give it back. Or he'd tell Jason once how to hold the gun then get furious if Jason didn't hold it exactly that way. He'd tell Jason, "Now you're not going to get the gun until you learn."

Jason would start crying, the signal for his mother to defend him. Mrs. J. would tear the gun or whatever it was away from her husband, give it back to Jason and tell him he could play with it any way he wanted. Jim J.'s usual reaction was to pour himself a whiskey and sit in front of television. The sight of his father's defeat and his sulking annoyance was very sweet to Jason.

What's wrong with this family scene? Everything.

Unwrapping a box or playing with a toy can be an exercise in logic and decision-making. By not encouraging Jason to solve his own problems, Jim seriously stunts his development in decision-making.

Jim J. may also be stunting Jason's intellectual development. According to studies by Professor Alfred L. Baldwin of New York University, children four to seven whose parents discuss decisions and their consequences with them have a higher I.Q. than children whose parents don't talk over judgments and decisions with them. Children whose parents don't care at all what they do or whose parents arbitrarily and dictatorily demand obedience also have poorer I.Q. ratings.

Jason's mother wants a perfect, picture clean house: no toys, no clothes lying about, everything in its place all the time.

If company's coming, she empties Jim's ashtray two or three times during each cigarette. Dust motes in the sun depress her so she keeps the blinds down during the day. She is repeatedly slapping Jason's hands telling him: "No! Don't touch! Bad! Bad!" She's afraid to let Jason try to climb because she's afraid he's going to break a piece of glass, or a cup, or get food on the carpet, or spill face powder on the bedspread. Jason sees his father accepting these house rules too.

Mrs. J.'s repeated victories over her husband may be teaching Jason something else, too—that it doesn't pay to be either a man or a husband. If Jason in high school expects his date to decide where to go or what to do, it's because as a child he learned it's the woman who makes all final decisions. Jason accepts a spot-free house as the best kind of house. When he marries, he may find his wife has different standards. She tolerates smeared drinking glasses, never empties ash trays and doesn't seem to see dirt. She and Jason will have trouble.

On an earlier page we offered guidelines for decision-making:

Don't run from making decisions.
Don't make them in panic.
Decide only what you're able to do.
Decide, then act.

These guidelines point to one master factor prominent behind most decisions: the decision-maker's image of himself.

Does he have a true self-image? Then he won't likely either overestimate or underestimate his ability to decide and act.

Is he an egophobe? Then he'll likely avoid making decisions. He will make them in desperation because he hasn't faced them when he was supposed to.

How does a person get a bad image of his decision-making ability? He gets them from the way his parents related to him when he first began making decisions during childhood.

Is he forever a victim of this relationship? Is he doomed to cringe from facing problems because his parents never trusted him to tie his own shoes?

No, he's not forever a victim. There are ways he can overcome a bad self-image. These same ways can improve the quality of his decisions and develop good habits of decision-making.

Quiz

1. What does decision-making have to do with becoming?
2. What is the chief difference between a judgement and a decision?
3. Give three reasons for the importance of small decisions.
4. Name the six steps before decision-making. Explain each step.
5. Name three steps to consider after decision-making. Explain each.
6. Give seven measurements testing any person you'd go to for advice.
7. What does *identification* mean?
8. How does *identification* differ from *imitation*?
9. How does self-esteem affect decision-making?
10. Why is decision-making man's chief means for *personal* becoming?

THE PREJUDICE AGAINST WOMAN DRIVERS

CHAPTER FOUR: COMMUNICATION

Communication–The Central Means for Social Becoming

Communication is a *transfer of meanings* and *feelings*. The transfer may be unconscious and accidental. A change of color in my face or a change of breathing may communicate fear or embarrassment or anger. I may communicate one thing verbally to a cop while my pulse is communicating something quite the opposite to a lie detector.

We communicate through touch, through facial expressions and gestures. We can use smoke puffs, drum beats, electric sparks, blinking mirrors or flashlights.

Since each of my decisions marks me somehow, my decisions also communicate who I am. Each leaves its print on my taste, my intellectual and moral character. Ultimately, these prints define my character and distinguish me from other persons. I can be recognized and identified by them.

Everything I choose to own, the friends I associate with, the clubs I belong to, whatever I organize or change or arrange says something about my SELF. Clothes, for example, can say:

I'm alive. I swing.
I'm in with this crowd. I *belong.*
I'm out of it.

OR THEY MAY SAY:

I don't care.
It's not important to me whether I belong or don't belong.

OR:

I'm ignorant.
I don't go out much.
My parents decide things for me.
I'm too busy to dress.
I've got more important things on my mind.

OR:

I hate authority.
I'm fighting my parents.
I'm fighting the school.

OR:

I'm shy.
I'm interested but afraid.
I like clothes but I don't want to be noticed.
I'm important. I do want very much to be noticed and liked.

The communication we're concerned with now in this book has four characteristics: It is (1) *deliberate* and (2) *conscious* communication. It is (3) *verbal* and (4) *oral* communication—words we mean talked face to face.

It's vitally important for me to know how to communicate myself because it is through communication I will be known and loved. When I communicate myself I extend my being. I become more.

Through communication I can avoid the calamity of being rejected. The most common rejection is to be lumped in a pile with others because of some superficial and accidental mark—like for being named Mary Smith, or for being white or black, or Polish or Jewish, or for being five feet six inches tall. If I can express myself to others so they can recognize me as a unique and distinctly different person, I will not be talked to as "just a Teenager." If I can define my conscience accurately, I can search for my particular sanctity, not the sanctity that fits somebody else totally different from me.

It's a mistake to think that communication by itself alone is magic and will solve all problems. A person can communicate murderous feelings, unreasonable hate, coldness and invincible prejudice, but is he and is the world necessarily better off for his ability to communicate?

Communication in itself is not necessarily good or truthful or heroic, no more than any other form of human action is necessarily good or heroic. Communication can be constructive or destructive. It is concerned with *becoming,* but the communicator has a choice: he can become a good man or a bad man.

The communication we will study is *constructive* communication. *Good communication is the simultaneous engagement of the Actual Self and the becoming of the Potential Self.* In this book we will refer to such communication as *dialog,* a term first made famous by the philosopher, Martin Buber.

Communication and the Listener

When a boy doesn't take seriously or believe what I say, I get mad and ruin the date.

Girl, 18, St. Louis

I myself am not a talker. I like to listen, analyze and then make an opinion. On this date, I did all the talking and learned nothing about this girl. This caused me to break with her.

Boy, 18, Milwaukee

I was with a very nice girl who liked to talk, but no matter what I said she paid no attention to it and kept on talking. For all I know she's still talking. I don't see her anymore.

Boy, 18, Denver

Any kind of sensitive talk is strongly affected by the attitude of the listener. Depending upon the listener, talk may be open or closed. An honest listener can make talk grow, develop and expand. A narrow-minded listener cramps it. A bad listener can make it falter, shrink, stop and die. Love draws talk out. Rejection deflates it and makes it collapse.

We all have different ways of coping with listener attitudes. A hostile listener makes some talkers try harder and makes others quit. Some talkers get angry when a listener interrupts, some get discouraged. Some blame the listener for yawning, some blame themselves.

The girl I took out was in a mood that everything you said was tear down or rib, therefore the rest of the night was quite silent and rather dreary.

Boy, 18, Milwaukee

Whenever I would try to talk to her she would either answer in a very crabby way or she wouldn't even answer. I finally blew up and left her.

Boy, 17, St. Louis

How do you talk to a boy who constantly complains?

Girl, 17, New York

It seemed I chose each topic which he disliked the most. He didn't care to hear about a certain girl, he didn't like her. A certain teacher he could not get along with, and he disliked my school because our basketball team had defeated his almost every time. I finally had to tell him to listen or not call back. He called back.

Girl, 18, New York

If we observe our talk over a period of time and with a variety of listeners, we will find we have certain habitual reactions to different kinds of listening. *We learned many of these reactions in early childhood.*

A child is born into the world marvelously simple. He isn't shy. He isn't afraid of appearing on television or on the stage of the Metropolitan Opera in New York. He can meet anybody, the principal, the Governor, the President of the United States. He isn't afraid of being messy; he smears mashed bananas in his hair. He can't pronounce a single word distinctly and intelligibly, but this doesn't keep him from trying. He knows no language but he'll talk to any foreigner who talks to him. He isn't at all nervous about his shortcomings—and who has more shortcomings than an infant?

But there are a few fears that a child knows very early in his life. He needs freedom to move, stretch, exert himself, so he will naturally fear what will keep him from moving, what will bind and restrict him. He is afraid of being abandoned, of not being wanted or accepted. He is afraid if he feels his mother's fear. These are some of his earliest fears. In the same way that they affect his earliest decision-making, they affect his earliest communication.

We're going to survey how our earliest communication skills grew. We will see that the success or failure of our communication ability was strongly affected by the earliest *listeners* in our life—our parents.

Communication Failure

High school seniors in St. Louis, New York City and Denver were asked to identify their most serious communication blocks.

GIRLS			BOYS		
Rank	Communication Block	Number of votes	Rank	Communication Block	Number of votes
1	Moodiness	165	1	Moodiness	118
2	Lack of common interests	126	2	Poor sense of humor	84
3	Poor sense of humor	125	3	Lack of common interests	81
4	Misunderstandings	85	4	Misunderstandings	43
5	Conceit—not listening	75	5	No place to go	34
6	Family interference	45	6	Lack of car	32

7	No place to go	39	7	Family interference	32
8	Age difference	14	8	Conceit—not listening	32
9	Race difference	14	9	Lack of money	20
10	Lack of car	13	10	Race difference	14
11	Lack of clothes	7	11	Age difference	6
12	Lack of money	6	12	Lack of clothes	2

In a later chapter, "Enemies of Becoming," we will see the chief "causes" leading to marriage breakups and divorce. You will see that some of the communication blocks which today spoil dates, will tomorrow spoil marriages.

Family interference affected communication on dates; *in-laws* affected communication during marriage.

Moodiness seriously hurts dating communication; temperamental conflict is a leading source of communication failure in marriage.

Moodiness

BOYS AND GIRLS BOTH RANKED MOODINESS AS THEIR #1 COMMUNICATION BARRIER

The moods which block communication are the sad, irritable moods. These sometimes abrupt, severe emotional changes affect openness. They close both talker and listener. We must cope with them as we cope with any other changes. It's selfish and unfair to make other persons suffer because of them.

Some moods are real, caused by fatigue, by hormone changes and subsequent changes in the body, by quarrels with parents, by self-pity. Some girls feel irritable or moody during menstrual periods.

Some moods are phony. The girl acts troubled or the boy pretends he is grappling with enormous black thoughts. Both want to be coaxed. They want pity or sympathy and will accept almost any kind of attention as a sign of care. Inevitably, unfortunately, the moods cheat them out of getting what they want most: to be loved, to be known, to love, to know. To have these they must communicate. But moodiness, real or fake, spoils communication more than any other cause in their life.

I was going out with this boy for about two years, not steady. I was dating others. Every time he gets moody he's very quiet. It's like

talking to a wall. The date ends up being boring, no fun or laughs and a fight results from all the tension. Maybe he should get it off his chest. But he lets us both suffer.

Girl, 17, Denver

The other person is moody and everything you say, you barely get an answer out of them. Then you feel that you did something to make them act this way, then you get mad and don't talk and everything goes wrong.

Boy, 17, St. Louis

Sometimes my boyfriend will just sit and will not talk to me and I don't know what to say because he won't talk, so I sit there for awhile waiting for him to get out of his mood. Then he gets mad at me for not talking and we have a big fight and the whole evening is ruined.

Girl, 18, St. Louis

Yes, I once was on a date where the guy was happy and carefree then all of a sudden got moody and depressed. It was very annoying and got me feeling the same way. I didn't know how to react.

Girl, 18, St. Louis

Whenever I'm out on a date with a boy who is moody I always feel uneasy, because I'm afraid that something I say might offend him, when I say nothing I feel like a poor date. So I avoid moody men, and try to find ones that love life, and are happy individuals.

Girl, 18, New York

I have a friend who gets moody and disturbed because of conflicts at home and he takes it out on you by ignoring what you say or being terrible to cope with. He will not openly discuss the matter with you—even though you might be a big help to him.

Girl, 18, St. Louis

Moodiness makes it hard for you to even talk to the other person without getting yelled at or something.

Boy, 17, New York

I wanted to go to a particular place and my date did not. We ended up not going where I wanted to which forced me into a mood, which caused the date to end in an argument.

Girl, 17, Denver

A boy I know is nice and sweet when he wants but when he doesn't he just sits and sulks, this bothers me because moods are something a person can control.

Girl, 18, Milwaukee

One evening instead of going out we stayed home and watched TV and played records. My date was in one of his moods. I'd tell him about something that happened at school or ask him questions and he would answer in short phrases. I had a hard time trying to stay in good humor after this. He left early.

Girl, 18, Denver

One night this boy I went out with was in a grumpy mood. I tried my best to start a conversation but got very little response. Finally I gave in and sank into sulkiness myself, we had a terrible time.

Girl, 17, Milwaukee

Lack of Common Interests

RANKED AS #2 COMMUNICATION BARRIER BY GIRLS
RANKED AS #3 COMMUNICATION BARRIER BY BOYS

Common interests can give important support to communication. They help us know and trust each other. They often supply a vocabulary we can share. We can exchange and appreciate common experiences. In other words, common interests give us a head-start in talk because we have something neutral and outside ourselves to talk about while we study each other and decide whether to communicate more deeply and personally. Common interests is the *money* of communication.

But the lack of common interests should not be a barrier to communication except for the unimaginative, the impersonal and persons who are already weak in communicating.

A limited vocabulary is itself a barrier to communication. Frequently ignorance of words is a sign of an ignorance of subjects. Persons who want to develop more interests do so by reading and developing an interest in words. But a big vocabulary won't guarantee communication. Some persons deliberately use unusual words to inflate their self-images, to hide their personal insecurity, to avoid dialog.

A person's genuine interests are a sign of his openness. Very few interests usually indicate a closed person, one therefore with a serious communication problem. A person with few interests is a poor listener.

> If you don't talk to your date you will never get a chance to find out her interests or find out if she's the type of girl you would like to know better.
>
> *Boy, 18, New York*

> I went out with a boy who just talked about cars and everything I didn't know too much about. I tried to ask interesting questions, but he just looked at me as if I was stupid. I didn't have any fun just listening to him talk all night.
>
> *Girl, 17, Milwaukee*

> I once dated a boy who was on a much lower level of intelligence than I was. We could never understand each other—he had no interest in my ideas and I had no interest in his. We didn't last long. I guess our minds and ideas just didn't coincide.
>
> *Girl, 18, New York*

> She will talk all evening about some hero she went with from college.
>
> *Boy, 17, Denver*

> You feel very uneasy and you and your date just aren't able to have a good time because you have different ideas and enjoy different forms of entertainment.
>
> *Girl, 18, Milwaukee*

A bad date is when you go out with someone who doesn't have the same interests. A date on which neither party knows what to say to the other and you know you won't have any fun. You find out that you don't have anything in common . . . for this reason I would not accept a blind date.

Girl, 17, St. Louis

I had a date with a girl who used all big words which I didn't understand and this made the whole night terrible.

Boy, 18, Denver

He played in a band and that's all he could talk about. Sure I like dances, but I didn't want to hear about every job he ever played at. Another date I had kept talking about what great times he's had with other girls, some of which are my girl friends.

Girl, 17, St. Louis

Communication is the most important, or should be condition for a date. If a person cannot talk about himself and the things that he sees and feels around him then it is impossible to share with him. There is nothing worse than spending an entire evening with someone who is or seems to be unfeeling, because he cannot communicate. He need not have the same interests as you and he need not have a penny to his name, but he must understand what it is to be living and wonder about it enough to discuss it.

Girl, 17, New York

Communication is the main part of the whole date. It can solely determine whether the date will be interesting or drab. No matter how nice a person you go with or how much a movie holds your interest, communication is very vital. Your interests don't have to be that alike because you can learn many new things about a subject you know.

Boy, 18, Denver

The guy may be bashful and not say anything or he talks about engineering and the girl might not know anything about it. They might just sit in the car or wherever they are and not say anything to each other.

Girl, 17, Milwaukee

Poor Sense of Humor

RANKED AS #3 COMMUNICATION BARRIER BY GIRLS
RANKED AS #2 COMMUNICATION BARRIER BY BOYS

A bad humor affects listening as well as the will to talk. It is a serious block to communication.

The so-called "sense" of humor is related to openness. A narrow-minded person has a poor sense of humor. A "wide" or broad-minded person has a more generous sense of humor. A person who lacks an open-minded humor is often an insecure person with B.S.I. He is oversensitive because he suspects people are ready to hurt him.

He has a poor sense of humor and you are afraid to say anything that might hurt his feelings.

Girl, 17, Denver

My boyfriend thought I was jealous over some girl and got angry with me for being possessive. I was just joking.

Girl, 18, New York

My boyfriend said something as a joke and he hit a weak spot. I pouted all night and he got mad at me. It certainly ruined the night.

Girl, 17, St. Louis

I said a couple of things in fun and he got real mad. He didn't know when I was kidding and when I wasn't.

Girl, 18, Milwaukee

I was in a very bad mood because of school and all he felt like was goofing off. It irritated me so much he took me home early.

Girl, 17, New York

Many times a sensitive person will be hurt by a seemingly cutting remark that was made jokingly.

Boy, 17, Denver

It spoils a date when you sit there like a fool waiting for him to laugh or when you sit and laugh and they just sit and stare in wonderment.

Girl, 17, Milwaukee

A girl that can't take a joke is a bad date. You can get a girl in the mood of the place you're in by joking around. If she has a warped mind you won't be able to get any place.

Boy, 18, New York

If you try to make a joke and the girl says nothing, you feel pretty discouraged.

Boy, 18, Milwaukee

I have had dates ruined because the fellow couldn't laugh at things that happened. He felt sorry for himself instead. The next time we went out and my best coat was stolen. He thought that was funny!!!

Girl, 18, New York

Misunderstandings

#4 COMMUNICATION BARRIER BY GIRLS AND BOYS

Misunderstandings cause communications to collapse. Ironically, misunderstandings are themselves frequently caused by a communication error or failure and the only way they can be properly corrected is through attempts to communicate again.

Misunderstandings also block communication in marriage. The unwillingness of persons to correct or clarify misunderstandings can harden and become a serious problem. One uncorrected misunderstanding creates another, and that still another. In the end, the original cause of the trouble is forgotten and lost. But two strangers are left who talk less and less, who know less and less of each other, who love less and less.

Daters ought to welcome misunderstandings. There is no better chance to practice communication during dating than during a misunderstanding. All one's habits of communication are suddenly exposed. Does he *trust* there was a misunderstanding or does he insist there was a deliberate attempt to hurt or deceive him? Can she admit she made a mistake or will she stubbornly keep saying all the fault was his?

> Someone else puts wrong ideas into your date's mind. Then she gets mad and won't discuss it.
>
> *Boy, 17, Denver*

> Something is misinterpreted by your date and gets bent out of shape.
>
> *Boy, 18, Milwaukee*

> One evening when my boyfriend called, I was lonely for I hadn't seen him in a long time. Instead of saying "I want to see you tonight," he said nothing, so I decided to say something and gave him all the hints in the world, at least I thought I did. Later I found out that he just didn't understand.
>
> *Girl, 17, Milwaukee*

> If one person uses words in a completely different way, a different meaning to the words, this can hurt a date.
>
> *Girl, 18, New York*

> I did not want to go out and get something to eat after we had gone to the movies. He got mad and thought I wanted to go home right away. I didn't tell him I thought he spent enough money on me when we went to the show. I didn't want to be greedy.
>
> *Girl, 17, Denver*

> I said something wrong. This took an hour to straighten out.
>
> *Girl, 18, Milwaukee*

> I made a remark which he thought was an insult, but it wasn't. The date was salvaged when I explained.
>
> *Girl, 17, New York*

> One time I misunderstood incorrectly and as a result a big argument came about. Things were then said that weren't meant. From then on we had doubts about each other as to what we thought the other person thought of our relations and reactions to one another.
>
> *Boy, 18, Milwaukee*

> One night I thought he was mad at me and he thought I was mad at him. Instead of talking the thing over we had a miserable evening for no reason at all.
>
> *Girl, 18, St. Louis*

Conceit–Not Listening

RANKED AS #5 COMMUNICATION BARRIER BY GIRLS
RANKED AS #8 COMMUNICATION BARRIER BY BOYS

Conceit is exaggerated self-esteem. It interferes with communication because the conceited person is not open. He doesn't expect anything interesting can be said by others so he doesn't listen.

He always wants to park. He never knows what to say. He treats you mean. He doesn't pay attention to you and what you say.

Girl, 17, Milwaukee

A bad date is someone who is too self-centered and doesn't even talk to you, much less worry whether you're having a good time or not. There has to be a sharing between two people of ideas. It doesn't matter whether you agree or not just as long as you are at least talking.

Girl, 18, Denver

I took out this stuck-up girl and I mean stuck-up! She didn't talk until later that night when we met some of her girl friends on the corner of Sixth and Mitchell, it was very uneasy. Didn't have a good time, if ya know what I mean.

Boy, 17, Milwaukee

A bad date is one that your date thinks she is doing you a favor by letting you take her out and she takes a snooty attitude about everything.

Boy, 18, St. Louis

A bad date is a guy who dresses like a slob, is a bore and still thinks he is God's greatest gift to women.

Girl, 18, New York

He thinks that he is the ultimate end and doesn't know you're alive.

Girl, 18, Milwaukee

I had no common interest with one boy. Lucky he talked about nothing but himself because there was nothing else to talk about.

Girl, 17, St. Louis

A fellow who was extremely good looking asked me for a date. I accepted and had the most horrible time. He was so self-centered and narrow minded that I just stopped trying to converse. He was also the only child and spoiled.

Girl, 18, Milwaukee

I call a date a flop if the boy tries to smart off all the time and can't even carry on a decent conversation.

Girl, 17, Milwaukee

Your date is conceited and thinks he is "Joe Cool" and trys to get what he can. He never continues the conversation or acts as if he's not enjoying himself. Even if he isn't interested he doesn't have to show it.

Girl, 18, New York

No Place To Go

RANKED AS #7 COMMUNICATION BARRIER BY GIRLS
RANKED AS #5 COMMUNICATION BARRIER BY BOYS

Having no place to go is a problem only to persons *who already have a communication problem.* These persons substitute action and motion for talk. They need to go to places to avoid facing their communication failures.

The girl doesn't like anything you do, you say, or suggest you do.
Boy, 18, New York

Usually you get tired of the same hangouts. You want to try something new but it seems like there is no place to go.
Boy, 17, Denver

There are many times when we go out on dates, we can't find anything to do. Either the shows are crammed or you've already seen the movie; bowling is fun but it is always crowded. This has caused us to be very moody.
Girl, 18, Milwaukee

I had a date, the car and we were already to go except we didn't know where to go. I don't like bowling, there wasn't any good show on, miniature golf wasn't open yet and the C.Y.O. is closed on Tuesday nights. We ended up sitting home with her mother watching TV.
Boy, 18, Milwaukee

Where is he taking me? I have a right to know. It's my reputation. If he doesn't show he cares about my reputation, I have to show I care.
Girl, 17, Denver

If you have no place to go you just sit and have small talk but pretty soon you'll run out of small talk and then you'll just sit there.
Girl, 18, Milwaukee

I think a bad date has nothing to do with where you go or what you do but what it does for you emotionally. By this I mean, if after a date you don't understand yourself or your date better, you may as well never have gone.
Boy, 18, New York

Family Interference

RANKED AS #7 COMMUNICATION BARRIER BY BOYS
RANKED AS #6 COMMUNICATION BARRIER BY GIRLS

Parents are a true cause for dating failure when they directly forbid a date or make it difficult to keep. Parents can spoil a date indirectly by causing bad moods or feelings of guilt which in turn affect communication.

I had a friend over for a day and my parents stayed with us *all* the time and did most of the talking and asking questions.

Girl, 17, Milwaukee

For no reason at all, my family really doesn't like the boy I'm going with. He tries very hard to have them get to like him. Usually my mom will say I'm not home when he calls, or says I can't go out tonight. We like each other very, very much. He says my parents won't break us up. But I'm afraid they will.

Girl, 18, New York

Once a guy asked me out, and then had nothing to say to me because he had a fight with his father. We both were mad, he because I tried to talk to him, and me because he wouldn't talk.

Girl, 17, Denver

I told my family I was going out with my boyfriend one afternoon (they said it was OK with them). My friend had to be to work at three o'clock that afternoon and he was coming at one-thirty so we didn't have that much time. Then my mother was in a crabby mood and so she didn't let me go until two o'clock. My friend was troubled and I was upset.

Girl, 18, St. Louis

I just had a real bad fight with my parents and the whole night was shot down.

Boy, 18, St. Louis

His mother told him that he was going to have to break up with me. He was worried and awfully upset. This made me feel bad to think I caused this argument between him and his mother and wondering whether or not I would have to break up with him.

Girl, 18, Milwaukee

Because of family interference, everytime I saw the person openly or secretly, I felt guilty and ashamed. I could not relax and therefore he could not relax either. We were just good friends but they wouldn't let us be.

Girl, 17, Denver

I just got home from work on Saturday at six o'clock. I was supposed to go out at quarter past six. My parents really got mad. They said I was never home. They told me I should leave and board at some rooming house. That night I just couldn't stop thinking about it so I was very quiet. It spoiled my date.

Girl, 19, St. Louis

I had family interference and moodiness both on one date. The boy I was with said his mother wanted us to break up. All night he was real moody and hardly talked to me until it was time to go home. Then I gave him the ring back and haven't gone out since.

Girl, 18, New York

It was raining and mom and dad had just had a big fight. My boyfriend picked me up. I was in a terrible mood. Everything he said, I took wrong. Before long there was a deafening silence . . .

Girl, 17, Denver

REBELLION

Rebellion is a refusal to accept. Frequently when an adolescent rebels, it's because his parents rebelled first.

Two immaturities are thus involved: the adolescent throws a tantrum because his parents threw a fit. The adolescent could not communicate with his parents because the parents were not first able to communicate with him.

REPRESSION: An unconscious defense mechanism. A person *represses* ideas, emotions or impulses when he avoids thinking about them. He dimly senses or suspects and fears the presence of these unwanted thoughts but will not face them. He is not aware that he's refusing to face them.

Repression is often confused with *suppression.* Suppression is a conscious act; repression is an unconscious act. They pair off just as imitation, a conscious act, differs from identification, an unconscious act.

At the same time, the parents themselves are beginning to move into a new stage of life and feel upset about it. They are *adolescent old people.* Their bodies may have long stopped growing but they are just now beginning to feel and see the changes. They may be aware of certain biochemical adjustments in their bodies and must now adapt psychologically to conform to the new conditions. They may see some of their oldest ideas being challanged by society, and they haven't yet figured out how to cope with all the changes. They go through changes of moods and attitudes; they feel unsure of the future. In a hundred different ways they rebel against the aging of their bodies. They feel their children are ungrateful and callous for not being able to see and help them meet these changes.

Parents will react to change according to their level of maturity. Mature parents will accept the idea that their sons and daughters will soon leave home. They tactfully avoid interfering with the normal *dynamics* of this period.

But some parents find this shift too hard. Their own aging process is forcing them to modify their own body images and adjust to new relationships. Their son's and daughter's so-called "independence" looks a lot like desertion and ingratitude. They may feel they are being rejected and react according to habit: they sulk, they pretend to be sick, they dangle bribes or toss off threats. They show their dislike for the new styles, the vulgar, coarse manners which to them are symbols of a world they don't want because it doesn't welcome them.

DYNAMICS: Emotional forces determining the pattern of feelings and behavior. These forces arise when drives and defenses interact.

REBELLION

DYNAMIC PSYCHIATRY: The study of emotional processes, and mental mechanisms. The study of the active changing factors in human behavior, such as the concepts of change, of evolution, of progression and regression. Descriptive pyschiatry is the descriptive study of clinical patterns, symptoms and classification. It is older, more static, more statistical than dynamic psychiatry.

While many parents accept aging with humor and grace, some rebel by a flurry of activities. They almost push their children out of the home. Adolescents who experience this may feel the same abandonment felt by an unwanted infant. They may react as they may have reacted in infancy with anger or self-pity. They may get even by doing something they know will hurt their parents. They may boast of drinking or using drugs. They may go housebusting. Few vandals realize that by breaking windows and splattering ink and eggs against walls they are repeating a form of communication they learned in their playpens as infants.

Perhaps the most violent act of deliquency is the impulsive pregnancy and the run-away marriage. Damages from these acts can't be washed off, painted over or patched. They make frightful scars on the parents and on the run-aways themselves, but worse, on their children. The scars last for life.

Lack of a Car

RANKED AS #6 COMMUNICATION BARRIER BY BOYS
RANKED AS #10 COMMUNICATION BARRIER BY GIRLS

One of my first dates the boy I was dating had no car. So his father drove us to the dance. The night never seemed to end. Later when the date had the use of the car the communication between us seemed more developed.

Girl, 17, Milwaukee

Not driving we always had to double with his brother. Without exception we both ended up in his brother's conversation. It seemed he needed a push to talk, much less think of something to say. But if he drove himself he wouldn't depend on his brother so much.

Girl, 18, St. Louis

I took out a girl four years younger than myself, whose interests as much as I could find out, were different. Besides that, I wasn't able to drive, because I can't afford insurance, though a car would be available (my brother's). The sum total was a crummy date. I think she was scared of my age and my interests, I know I feared offending her taste with talk of my interests, and I was greatly embarrassed at bus travel—I think no car really did it, for at least I would have "talked up" then, as I have done earlier with other girls.

Boy, 18, Denver

THE CAR AND COMMUNICATION

The car affects communication in these several ways: the car is *transport* to a place for communicating. It gives freedom and mobility. It is a *site* for communication. It offers privacy, quiet and shelter. It is a covered couch on wheels. The car affects the owner's *image*, both his *self-image* and the *image he projects* to others.

The car is something to talk about so it is *content* for communication. The car's driver uses it as an *instrument* to talk *through*. It can "say" all the things clothes can say.

Approximately half (47 percent) of the teenagers asked by the Purdue University Poll said that once a week, or more, they spent an evening riding around in a car with friends. The Purdue University Opinion Panel asked: "Do you think a car helps a boy get a date?" Seventy-eight percent of the boys and 67 percent of the girls said *yes*, a car does help get dates. Seventeen percent of the boys and 28 percent of the girls said it didn't.

Do you own a car?

	Boys	Girls
Yes	32	7
No	65	90

Purdue Opinion Panel

Girls object that riding around in a car is a *boy's* notion of fun, not their's. For them it's a danger, a worry, a threat, a source of trouble and a cause for boredom.

If you happen to pick a guy that seems like a lot of fun, but when you're alone is like a clam and never talks or just grunts, and you don't go anywhere, just ride around.

Girl, 17, Denver

I like to have an idea of what we'll be doing. I'm not crazy about just driving around. I went out with a racing maniac once and spent the night hanging onto the door strap with his fast driving. If he wants a thrill-ride let him take me to a carnival.

Girl, 18, St. Louis

All we did was drive around and talk. Drive around and *talk*! That's a date?

Girl, 17, St. Louis

I can't stand the boy who doesn't talk or try to further the conversation in any way. Doesn't even want to do anything but "ride around."

Girl, 18, Milwaukee

But to a boy the car's power is his power. The waving plume on the high whip antenna, the juke box lights hung on the grill, the jet exhaust pipes glowing like rockets behind them, the radio with its antenna slashing the air and its volume pushed to the floorboard, all these are symbols. They are supposed to tell others what a smoking cannon he is. He's a tiger is what they say.

He wishes the girl next to him weren't such a log. He tries to impress her. He bites his wheels into the road when he accelerates. They squeal and scream when he commands speed.

What the girl doesn't realize is that the car is the boy's means of communicating. He doesn't talk with his date. The car talks for him. He uses his car to communicate feelings and dreams. But she can't understand because driving, like music or painting is a special, limited language. It isn't clear.

Every cut in and out of traffic, every turn represents a decision. Everytime he cuts out of his lane or passes another car he's showing how he rebels against authority.

To make sure she appreciates him, he comments on other drivers and underlines each comment with searing horn blasts. "Look at that jerk!" *HONK.* "Wouldn't you know—a woman driver!" *HONK.* "What does he think he's doing—*Signal next time buster!*" "Where did he learn to drive?"

It's bad to go out with a person if all you do on a date is ride around in what he thinks is a "cool car."

Girl, 17, Denver

The non-driver doesn't realize how much joy the boy gets from mere driving, that he likes the feeling of power it gives him. He can control his destiny by the way he steers. He can go fast or slow. He can turn. The feel of the road is soothing. The tiny vibrations that the road passes up to him are relaxing. He can't think of a better evening than having a girl companion and tooling over the roads.

What the boy doesn't realize is that being driven around is dull business for the girl. She isn't busy like the boy is. She's being carried around. The route is a little different, but she can get the same effect by riding on the bus. She's seen the streets. He's going too fast for her to see the shops. She has no sense of knowing whether she is five miles or two miles closer to any destination. When the bus carries her, at least she knows where she's going. He can't talk with her because his mind has to be

on the road. If his mind isn't on the road she's afraid of an accident and *her* mind has to be on the road. When he's in a bad mood he shows it by the way he drives.

> I have this buddy. He's in the Army now. He was going with this girl for two years. Then she broke it off. No warning. She didn't want to go steady anymore. He used to pick me up and I sat with him while he made right angle turns at 50 miles an hour. He didn't have to talk. I knew what he was saying.
>
> *Boy, 18, St. Louis*

Her fun is *being his date*. If that's no fun, she's having a bad evening.

She tunes her mind into the radio because at least it's something. HER SONG comes on and she turns it up. She wants it to talk for her, and tell him what she's feeling. She closes her eyes, rocks, juts out her chin, nods her head to the music. The song is supposed to say *I'm sad, talk to me,* or *I'm happy to be out of the house*, or *life is a bore*.

He asks her something and she doesn't hear. The radio's cut him out. He misses her message because music isn't a clear language either.

Lack of Money

RANKED AS #9 COMMUNICATION BARRIER BY BOYS
RANKED AS #12 COMMUNICATION BARRIER BY GIRLS

Money influences decision-making. It may decide whether a boy can ask a girl to a dinner and film, or a 19¢ hamburger. A boy's buying power affects his self-image. In this sense, the lack of money may be a communication barrier. But, frequently, a boy may spend money to cover his communication fears and failures.

> Some guys just ride around and want what they can't get and aren't willing to spend any money whatsoever.
>
> *Girl, 17, Milwaukee*

> You spend money on a girl and she keeps asking you to spend more (if this is accompanied by her talking about other guys it's worse).
>
> *Boy, 17, Denver*

> All I expect is a little something to eat, not a very expensive place.
>
> *Girl, 18, New York*

> The worst kind of date is one which I'm tied from "going all-out" by limited funds (I have always been ill-at-ease on dates because I must regulate spending so tightly).
>
> *Boy, 18, St. Louis*

A bad date is one who thinks the only way to have fun is to spend money on entertaining the other person. The main idea is to communicate with each other and to learn—with an unfeeling person with no viewpoint you cannot.

Girl, 17, St. Louis

I go out to have fun and a good time. It doesn't have to be expensive. I'm happy if my date isn't worried if he has enough money or gas. You shouldn't go to extremes moneywise on dates.

Girl, 17, Milwaukee

Lack of money seems to hinder me most on dates. This past prom is a classic example. After the dance at the War Memorial we went out to eat at the Stagecoach Inn. But because of lack of money we couldn't eat the things we wanted but had to settle for something at a lower price. That made me feel pretty cheap but what could I do? Other times I have to turn down going places with my girl because I haven't any money. I'm sick of always borrowing money from my friends or parents. It's really a sickening state. I have a job but the money goes as fast as I make it.

Boy, 18, St. Louis

Age Difference

RANKED AS #8 COMMUNICATION BARRIER BY GIRLS
RANKED AS #11 COMMUNICATION BARRIER BY BOYS

Senior girls often date college boys. Senior boys frequently date sophomore (that is, younger) girls. Calendar age of itself, is no special barrier to communication. But a girl's image of the older boy's "sophistication" along with an image of her own inadequacy may stifle her attempt to communicate.

The only date that I've had ruined because of a lack of communication was a double date in which my boyfriend's friend brought a girl that was much more wild than I was at the time. She and her date carried on and I found it impossible to respect these two and was mad at my boyfriend for not telling them to wise-up or get out of his car.

Girl, 18, Denver

When the boy is very intellectual and in college and when I'm still in high school—communication is revolving around education—self or school education.

Girl, 17, Milwaukee

Right now I'm going with a guy twenty-three. Now, of course he doesn't want to hear what happened in homemaking class or typing because he's not that interested. Therefore some nights can be pretty dull.

Girl, 17, St. Louis

I went on a blind date once, and I found out the guy was twenty-one years old. I didn't know this before otherwise I wouldn't have gone. We didn't have much to say to each other and it was a real drag for us both, we had a miserable time.

Girl, 17, New York

The girl was only a sophomore and I was a senior. She was real good looking and didn't look like a sophomore at all. Because of this I thought she would act older. But I was wrong. It was one of the worst or maybe *the* worst date I was ever on. I was lucky I was doubling with a friend or it would have been a real failure.

Boy, 18, New York

We went to a movie and we doubled with a friend of mine. She was quite young, fifteen years old, and she was my blind date. I don't know whose fault it really was, but she wouldn't say a word, and when I did ask her opinion of something she answered with a quick yes or no!

Boy, 18, Milwaukee

A boy was five years older and lived in another city, he was real nice but I couldn't understand half of what he was talking about. He was a college graduate and I was a junior in high school.

Girl, 17, Denver

Race Difference

RANKED AS #9 COMMUNICATION BARRIER BY GIRLS
RANKED AS #10 COMMUNICATION BARRIER BY BOYS

Race difference is, in itself, not a barrier to communication. False, prejudiced images set up the barriers. Twenty-eight students said race differences had blocked communication during a date, but not one student explained how. In schools where interracial dating is not rare, communication problems will not appear as racial differences but as "moodiness," "no place to go," and "family interference."

Lack of Clothes

RANKED AS #11 COMMUNICATION BARRIER BY GIRLS
RANKED AS #12 COMMUNICATION BARRIER BY BOYS

Lack of proper clothes may affect a person's self-image and so interfere with communication. Only eight students—

six girls and two boys, complained that clothes had been a dating communication problem.

> Like he didn't tell me his mother was going to be there and I wasn't ready. My hair was a mess and I felt like a grub. I didn't say a word all night.
>
> *Girl, 18, New York*

Communication: How It Starts

Human communication grows up through stages: from crying to facial expressions and words, from touch to gesture, from receiving to sending to exchanging, from considering self alone to considering self, family and community. But to reach from stage to stage it must be exercised.

Communication failure during the dating years usually points to some disturbance at one of the earlier stages of development. One such disturbance could come if a child skipped one of his development stages or didn't get enough encouragement or exercise during the stage.

A parent who interrupts a child and finishes what he wanted to say or a parent who guesses ahead of the child or who doesn't answer or who listens to TV instead of to her children, interferes with his development in the same way that the parent who won't let her children solve their own problems interferes with their decision-making development.

Children have counting games, for example, ball bouncing and rope skipping games. These games help develop the important sense of timing and rhythm, a sense of *when to come in*. Some people never know *when* to speak. They doubt whether they should ask a question now or save it until later. They fumble jokes. They misquote or misinterpret. They speak off the point of the conversation. In general they lack timing or lack confidence in it.

> I have trouble communicating with a lot of people. I guess I don't understand others or they don't understand me.
>
> *Boy, 17, St. Louis*

> Everything we talked about didn't go over and finally when we did start talking it was about all the things we hated about each other and we got mad and both stopped talking, so I took her home.
>
> *Boy, 18, Milwaukee*

LANGUAGE COMMUNICATION

The infant is born equipped to make noise and he does it naturally. No one teaches him. During the first few months after birth, the baby's sounds of comfort and distress are wrung from him by bodily states he is helpless to control.

Crying is the infant's first attempt at oral or tonal communication. His parents have to learn to distinguish cries which mean "food" from those which mean "hurt," "discomfort," "fear." Thus parents are the child's first listeners. They are the most important listeners in his life because they help him form the basic habits of communication.

In the third or fourth month, the child discovers the magic of babbling. He learns control over the making of sounds. Babbling is a new achievement and he enjoys playing with it. He repeats his sounds, varies them, runs them off in patterns. It's baby jazz.

What's happening is that the baby is laying the groundwork for learning his first language, the language spoken by his immediate family. At this period of his development he can gurgle sounds deep in his throat or squeeze them out of his nose. His pronunciations are accidental and he doesn't know what he's saying but he can pronounce *tzchcin* or *hmhummuch* or *spanferkel* or *Worchester* or *Gaulois*.

As parents learn to identify cries, they name them for the child. Gradually, he learns his first words imitating his parents and using the muscles he has developed from sucking and feeding: *mama, wawa* (water).

Compared with language, crying is inefficient and unreliable, but the child learns to trust it because it gets results.

Our attitudes toward talking and listening are formed at home the same way other images are formed there, through the good or bad example of parents, through *identification* with them and through *direct instruction* from them. Our habits of communication are especially formed by the listeners who surround us at home.

During his first three months, a child born deaf will make the same sounds as any normal child. It is after this time that deafness will affect his speech because it prevents him from imitating sound. He doesn't hear anything to imitate.

A child is not born with the complete power to imitate. He has to learn it. A child will imitate a human sound only if he is encouraged to do so. The parent *must smile* and *make sounds* himself. A British language psychologist proves this by an interesting experiment. He asked a young father to stand at the foot of a crib and merely look at his five-week old son.

The father was instructed to make no sound and to keep his face expressionless. After six minutes, the child made only two or three sounds. The father was led out, then brought back later. This time he smiled, waved and said "hello." In the same period of six minutes, the child made as many as thirty sounds. The experiment was repeated with many other parents and children, always with similar results.

An uninterrupted relationship with an affectionate adult is the most important condition for early language learning.

Anna was born in secret out of marriage. Her mother, ashamed, never reported her birth. She hid Anna in a locked attic room. She left her completely alone, saw her only to feed her. By hitting her, she trained her not to cry.

As far as is known, Anna never had friendly contact with another human being. When she was discovered at the age of six she couldn't talk or walk. She appeared to be deaf and blind too. She had never been trained to use a toilet so couldn't keep herself clean.

Anna was taken from her mother and an attempt made to help her develop properly as a human being. It took months before she learned to respond to attention and months before her face began to reflect any feeling. By the time she died at the age of ten years six months, Anna had developed to the level of a normal two or three year old child. She had learned habits of cleanliness. She played with a doll and blocks. She could use some words and phrases and was attempting to talk.

Dr. Aaron Rutledge

The mother-child exchange of sounds is the child's first experience with oral communication. To make this experience successful, the listener must be interested, must show pleasure, enthusiasm and encouragement. The listener must be patient and not nervously hurry the child. The listener must be confident the child can respond. Eventually the child learns to be confident too.

This is how the child learns the language of his family. As he grows, he discriminates sounds he likes from angry and harsh sounds that scare him. He recognizes sounds of surprise and panic and fright. He learns these meanings long before he is able to pronounce the words that carry them. He learns strain and tension in the family. He learns love and care, peace and relaxation.

Most important, the child learns to recognize *attentiveness, care* and *interest*. He can tell from his mother's voice when she is distracted or busy with something else. He learns to detect when she doesn't mean what she's saying. He learns to know when she's lying. He can tell when she's sincere, too; when she's excited, happy, feeling good; when she's proud of him and when she's disappointed. He can learn all these things before he learns how to talk.

It is foolish to disguise unpleasant news by spelling it over the heads of children. They already have picked up the meaning from attitudes, gestures, facial expressions and tones of voice. They recognize symbols—a mother's coat and hat mean she's going out.

The relationship the child has with his parents in his early years of sound and language learning will affect his attitudes toward communication for the rest of his life. It is the source of his success or failure to communicate during dating and marriage.

Crying, the child's earliest experience with tonal and oral communication, gets results and gets attention. But there comes a time when the little boy is told to stop crying. Instead of telling someone else he's been hurt, the boy is now expected to report hurt only when it's very serious. He is expected to "fight his own battles."

He's told "only girls cry" thereby being falsely taught that girls are weak because they cry—and cry because they're weak, and that boys are good and brave for not crying. Later in life he will feel guilty, embarrassed and ashamed for crying or otherwise expressing honest emotion.

Here is one valuable means of communication cut off for a boy. One which can have much to do with his communication development. The boy who's punished for crying gains different experiences from girls who are permitted to cry. Girls learn to verbalize difficulties. They learn to talk them over with others. Boys see they are expected to act, not talk. They may even distrust men who talk a lot. They are expected to depend less on others, more on themselves. This independence has advantages for the kind of life a boy may have to develop, but it has disadvantages so far as communication in a modern highly organized society is concerned.

Crying, besides being a means for communication, is also a release of nervous tension and emotions. Boys are denied this form of release and may find it hard to talk out their troubles later. They may find it hard to reveal their feelings.

Mr. Dade J. is a bullying, table-pounding talker who ends each argument with his wife by spitting out, "Women! Women are so stupid." Mrs. Dade J. knows her husband has to win or he'll sulk the rest of the day.

Dade Jr. identifies with his father. When he's grown up and dating, the girls he likes to date best will be those who are compliant and agree with him. He'll have nothing to do with girls who put out strong views of their own.

The Dade J's have a little girl who identifies with her mother. The little girl concludes the best way to get along with a man is to let him have his way. Even as a child she will try to please her father by letting him know it all, by asking him questions, by depending on his decisions even when she can decide for herself. In school, she may wonder what's the use of studying since, if she gets too smart, boys won't like her. Later, during dating, this girl will expect the boy to lead the conversation and make all decisions.

Suppose Dade J. bellows at his son the same way he does at his wife. Dade, Junior may find that the trick to getting along with his father is to behave as his mother does. He will get very little practice at self-assertion and this will show up during his dating years. Or suppose Dade J.'s daughter pities her mother and makes up her mind no man will ever treat her this way when she grows up. She may sharpen her wits and try them against her father, try to catch him in mistakes. Later in life she may spend date after date proving to boys she's just as smart as, if not smarter, than they are.

Children imitate what they see at home. They may see father come home in the evening, talk a bit, play awhile, read, watch television. He may be tired of talking all day at work. He offers no example of lively, interesting conversation. But mother meanwhile has been home talking with the children, talking with neighbors, talking on the telephone. When father comes home, she talks with him. It's not surprising, therefore, that girls talk on the telephone more often and longer than boys. Twice as many girls as boys spend between fifteen minutes to one hour a day on the phone. Three times as many girls as boys spend more than an hour a day on the phone.

Whether a girl thinks it's manly for a boy to talk or be silent will ultimately depend on how she felt as a child toward her mother's and father's talk. If a boy believes he should dominate the conversation and believes all girls' talk is empty prattle, his prejudices may be traced back to his earliest years with his family.

> It's bad when a girl just sits there and doesn't say a word, and whenever you try to start a conversation she just agrees. She sits there like a piece of petrified wood, it makes me feel uneasy and afraid to say anything. A successful date to me is when the girl

keeps a good conversation going and doesn't always agree with everything you say, she has a good sense of humor and she is relaxed.

Boy, 18, Milwaukee

On this date, the girl just agreed to everything I said. It was no fun talking because I knew she would agree. She had no brainpower of her own.

Boy, 17, St. Louis

He wanted to see a movie I didn't care for. So we went. Then he kept saying "I know you didn't like it." Well I said: "I didn't, but that's OK." He didn't like that. We didn't agree on anything else either.

Girl, 17, Milwaukee

It can be an evening of harmonious discussion or even a heated debate but I like a date who talks.

Boy, 18, St. Louis

As the child collects more needs, words themselves become a need. He has to have them to point more precisely to what he wants. The more complex and mysterious his needs, the more precise must be the pointer. He will study language and art and prayer and music and mathematics as different means of expressing his deepest basic needs. Yet all his life, at each discovery of his lonely, separate, individual condition, he will "cry" when he longs for nearness. The more words he knows the better can he identify his differences and define his individuality, the better will he be able to reveal himself to others.

During infancy, the ability to play with sounds is a source of joy to the child. It is also a sign of his growth. As he grows older, as he learns words, he needs a *listener* to complete his growth. He needs a *watcher* if he is to make gestures or facial expressions. Talking to himself and making faces in the dark is a waste of time. The *pleasure of talk depends on the listener's pleasure.*

If a child finds no encouraging listeners at home, his communication later will certainly be affected. If his listeners tell him to shut up; if they tease him because he lisps or mispronounces words or stammers; if they correct his speech too often or too much; if his teeth are so bad he's self-conscious of them—in these ways and in many more a child's environment of listeners can keep him from learning habits of talk. He's particularly discouraged if nobody cares enough to listen to him. Disappointment at not being heard will trouble him all his life.

Children will also copy the self-esteem of the parent they identify with. The worth a person puts on himself is tied tight to his mode of expression. Dad may never voice an opinion because he doesn't think much of his ideas. Nobody does anything when he gives orders. Nobody listens when Mom talks. Without the slightest justification in fact, a child may gain the attitude he has nothing worth saying to anybody.

The student has . . . interacted with his family tens of thousands of times. He thus has been exposed to almost innumerable signs as to whether others are interested in what he has to say: the stifled or open yawn when he speaks, the interruption or changing of the subject, the look of distractedness when he expresses an opinion; or, on the other hand, the light of interest when he presents his views, the responses appropriate to his comment, the encouragement to continue, the request for his opinion on a subject which others have initiated . . .

Morris Rosenberg

Suppose Dad's eyes glaze over and he sinks into a stupor while Mom yammers about her day downtown buying curtains for the attic windows. Dad, here, isn't the only non-listener. A good listener would hear her own dullness. Mom's talk is more a nervous release than a serious attempt to say something worth giving attention to.

Parents who don't listen to their children very likely don't listen to each other. The child's first experiences with rejection during infancy are repeated when he sees his father and mother not listening to each other.

Not listening is a form of rejection. It leaves serious scars on the self-image. The child feels the non-listener doesn't care and has deserted him. He feels what he has to say must be worthless or wrong. He sees his father zoom in on the newspaper, his mother's ear locked to the telephone. His parents don't talk to each other and they don't talk or listen to him. In countless U.S. homes, blue TV-lighted rooms are lethal chambers destroying family conversation.

I haven't talked seriously with anyone around here in ten years. My father is not the type who sits and listens. He sits and tells you—when he bothers. I sit down to talk to my father, and he falls asleep. If I take a problem to him, he immediately jumps into a stand and tells me what to do. That does me no good. I'm old enough for a discussion, not an ultimatum. One day, I came home to tell my folks I had just gotten an "A" in Advanced Placement physics. My father and I got involved in something the minute I walked in the door, and it ended with him telling me I'm not old enough to have serious opinions about anything. I walked out of the house. I never did tell him about the "A". I can't win an argument with my father. So I duck him or become indifferent when he's around. My mother says I must respect him. I want to. But he has to have some respect for me, too. With the Navy next year and then college, if I'm lucky, this will be my last year at home.

Warren C., 18

My father works one hundred hours a day. He brings work home every evening. He goes into his office on Sunday. A month ago, he went into the backyard and shot some baskets with me. He thought he was a hero because he spent fifteen minutes there. I couldn't wait until he left. I literally have no idea who he is. How could I? We have rarely discussed anything meaningful. We used to talk at dinner at least. Then, he bought a small television set and put that in the kitchen. We were forbidden—forbidden—to talk. I tried once or twice and quit. I got sore. I wanted to talk. I got a job as stockboy in a department store. It keeps me out until after dinner is finished at home. Now, I go three or four days without seeing my father. I tell him he works too hard. He says he has to work hard to build a business for us. By the time the business is paying for itself, who'll need it anyway. I used to wonder why he was hiding from me. Now, I don't care.

Jerry K., 17

Quiz

1. Why do we say a person's parents are the most important listeners in his life?
2. Is there any relationship between communication blocks during dating and communication blocks during marriage? Explain.
3. What is "conceit"?
4. What is the difference between suppress and repress?
5. Can race differences block communication? Explain your answer.

Martin Buber

BORN: FEB. 8, 1873, VIENNA.
DIED: JUNE 13, 1965, JERUSALEM.

Five hundred years ago, men could hold contradictory opinions and ways of life because they were separated from each other by vast stretches of land and water which had to be crossed on foot or horseback or on wooden ships blown by the wind. Chinese, Africans, Germans, Indians, Spanish and Romans could worship different gods, practice different marriage and family customs, support different forms of rule and power.

They could do everything in privacy surrounded by thick walls of time and distance. Travelers seeking to penetrate those walls were in danger of being assaulted by hunger, thirst and disease or attacked by robbers.

Customs and ideologies conflicted only when journeying soldiers or merchants brought them together. The typical way to deal with dissenters was to quash them, make them slaves or throw them out of the country.

Our world today is bigger, faster and more crowded than the worlds of Thomas Aquinas and Aristotle. Television, communication satellites, computer calculators, jet and rocket transportation are bringing people of the world closer and closer together. Travel is no longer threatened by highway robbers or pirates. Never, never before as today are people so conscious they have different Gods and gods, different marriages and families, different governments, different taxes, different diets, different weapons, different alphabets. Banishment and eviction for dissenters is no longer practical. Execution for disbelievers is illegal. Slavery is unpopular. Meanwhile, the world is filling up. How will we ever be able to live together without the walls of time and space? There are fewer and fewer islands.

When Greeks and medieval philosophers considered how man got along with man it was usually to study his ethics and morals, his justice and charity. It was more important for early philosophers to understand how man related with God and nature. But the modern philosopher's problem is to explain how man relates to man.

Two vicious world wars and some smaller wars have taught us a serious lesson. It is this: *when men who disagree stop talking they start hitting each other.*

The modern thinker, if he will help men survive, must try to help them keep talking not only with God in prayer but with each other.

Such a thinker was Martin Buber, the philosopher of interpersonal communication.

All men have access to God, but each man has a different access. Mankind's great chance lies precisely in the unlikeness of men, in the unlikeness of their qualities and inclinations. God's all-inclusiveness manifests itself in the infinite multiplicity of the ways that lead to him, each of which is open to one man.

Martin Buber

Martin Buber's thinking was inspired by the Old Testament, modern psychology and plain, ordinary common sense. He is regarded as a pioneer bridge-builder between Judaism and Christianity, particularly between Judaism and Roman Catholicism.

In Jerusalem, Buber lived simply in an old stone house with red tile roof and garden. The house had been abandoned by an Arab-Christian family shortly before Israel's war for independence broke out in 1948. His study and bedroom were on the ground floor. The furniture was old and dark and heavy and the house was stuffed with books.

When a caller telephoned, he would hear a quick, sharp; "Buber." When receiving visitors in his study, Dr. Buber would be without tie, in shirt sleeves and slippers behind his desk.

In conversation he would rest his interlocking fingers on his paunch, then lift them to his forehead as he considered his response. He thought over his words carefully. His voice, when he finally answered, would be slow, low. To many it seemed tinged with melancholy. He disliked general questions and frequently prompted his questioner to "be specific." His letters, similarly, were brief and "specific."

In Buber's view, the Bible was neither an infallible guide to human conduct nor a mere collection of legends. It was, he said, a dialog between God and Israel, with Israel as *Thou* and God as *I*. This interpretation of the Bible and reports that he was casual in his observance of Talmudic law caused many Orthodox Jews to suspect Buber as a heretic. Some Reform Jews also criticized him because, they said, he drew many of his ideas from the Hasidim, a sect they despised.

Buber was Professor of Comparative Religion at the University of Frankfurt, Germany, from 1923 to 1933. The Nazis locked him out and canceled his professorship. Buber kept speaking, however. Once in Berlin, he lectured to an audience which bristled with two hundred menacing Storm Troopers and Gestapo. Hitler silenced him in 1935. Soon after, Buber was forced to escape from Germany and flee to Palestine. He was Professor of Social Philosophy at the Hebrew University until his retirement in 1951. Though he had suffered under the Nazis, Buber opposed Israel's execution of Adolf Eichmann, the Nazi, for crimes against the Jews.

"For such crimes," he said, "there is no penalty."

Many Jews were dismayed because Buber asked for an Arab-Israeli dialog. Buber told his critics: "The love of God is unreal unless it is crowned with love of one's fellow man."

Martin Buber wrote more than seven hundred books and papers. But he is most universally known for a book published over thirty years ago named "Ich und Du"—I and Thou. In this book, Buber introduced his concept of dialog as man's chief way to fullness and authenticity.

I–Thou vs. I–It

Martin Buber said man has two kinds of relationships: *I-It* and *I-Thou*.

The "I" stands for my total, unmasked, unfaked SELF. "Thou" stands for *you*, your total, unmasked and true SELF.

An I-Thou meeting, whether between man and God or man and man, whether in love or hate, is personal and perfect. It is a true human meeting engaging man's finest qualities.

Dialog doesn't necessarily mean agreement nor does it mean reflex disagreement.

> I accept this person, the personal bearer of a conviction, in his definite being, out of which his conviction has grown—even though I must try to show, bit by bit, the wrongness of this very conviction. I affirm the person I struggle with, I struggle with him as his partner, I confirm him as a creature and as a creation.
>
> Martin Buber

In the I-It relationship *I* treat another person as a thing, an object, as something to be used, as something to be beaten. The *I-It* relationship is impersonal and imperfect. If I fake or mask myself when I'm with someone, or if I let someone relate to a false image of me, I am engaging in an *I-It* relationship. If I relate to someone else whose SELF is faked or masked, or if I relate to my own false image of that person I achieve no I-Thou meeting.

In religious matters, the I-It is expressed when man regards God as out of reach, or builds theological systems in which God is remote, or considers God as something to be used.

DR. MARTIN BUBER

Buber insisted that God was personal, the "eternal Thou," not an It.

In the I-Thou relationship between man and man, two persons face and accept each other genuinely as full human persons. One person does not use the other as a stepping stone to meet someone else, nor as a tool to serve himself, nor as a mirror to reflect him. At the same time, each person accepts himself for what he truly is, no more, no less. He does not fake—present a false-fronted self. I-It is Master and slave. I-Thou is Man and Man. The I-Thou meeting, because it rips off pretension, permits man to *meet himself.*

Buber contended that man could realize an intimate relationship with God through intimate communication with his fellow man. This intimacy could be achieved in a "dialog" between man and God. Buber developed this concept in his work, "I and Thou," in which man is "I" and God "Thou." He called this I-Thou meeting *dialog.*

In this book we adopt two definitions of dialog constructed by Reuel Howe, author of THE MEANING OF DIALOG (Seabury, 1963): (1) *Dialog is the address and response between two or more persons in which the being and truth of each is confronted by the being and truth of the other.* (2) *Dialog is the meeting of meanings.*

The Problem of Seeming

In Martin Buber's opinion, man's central problem rises out of his way of learning and knowing. It is the problem of *seeming vs. being.* Because man abstracts his knowledge from sense observations, he must always be on guard not to mistake *appearance for reality.*

> To yield to *seeming* is man's essential cowardice, to resist it is his essential courage.
>
> Martin Buber

Socrates, Plato, Aristotle and Aquinas, in fact all philosophers recognized and struggled with this the *problem of seeming.* But they wondered usually how it affected man's relationship with God or how it affected thinking itself. Buber wondered how it affected communication.

The Greeks would note that an oar in water looked bent. "My eyes tell me the oar is bent," they said. "But my reason tells me it is straight. Which is right?—my sense of sight or my reason? If my eyes are right, how can I trust my reason? If my eyes are wrong, how do I know whether they ever tell me the truth? If eyes lie, how can I trust my ears, my nose, my tongue, my fingers?"

Buber trusted his senses and his reason but had doubts about whether men could be trusted. Some men lived behind appearances and depended on images they were able to project to others. They were worse than liars, Buber said. They made their very existence a lie. They were hollow men, stage-set men.

> Let us . . . imagine two men, whose life is dominated by appearance, sitting and talking together. Call them Peter and Paul. Let us list the different configurations which are involved. First, there is Peter as he wishes to appear to Paul, and Paul as he wishes to appear to Peter. Then there is Peter as he really appears to Paul—that is, Paul's image of Peter, which in general does not in the least coincide with what Peter wishes Paul to see; and similarly there is the reverse situation. Further, there is Peter as he appears to himself, and Paul as he appears to himself. Lastly, there are the bodily Peter and the bodily Paul. Two living beings and six ghostly appearances, which mingle in many ways in the conversation between the two. Where is there room for any genuine inter-human life?
>
> Martin Buber

The First Principle of Dialog

The more open I am, the more impressions will I receive. The more open I am, the more will I express.

This is the first principle of *becoming more: Be open.*

When I am open to being taught, when I listen to a reader or singer or teacher, I *become* what is read or sung or taught.

There are two becomings, then, mine and the singer's or talker's or teacher's.

Likewise, there are two becomings when *I* talk: the open listener's and mine.

A CONFERENCE OF GHOSTS

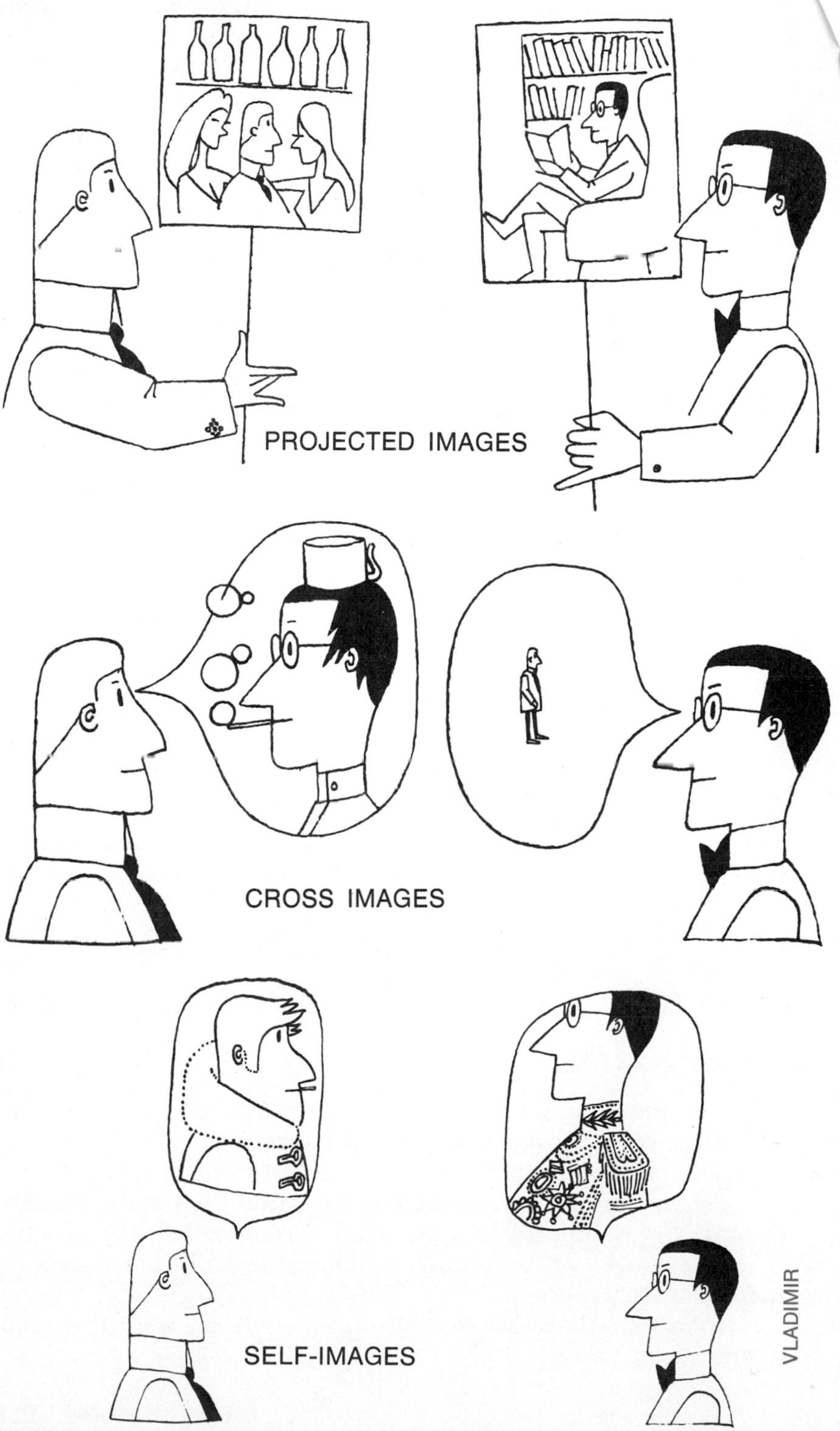

PROJECTED IMAGES

CROSS IMAGES

SELF-IMAGES

VLADIMIR

What does "be open" mean? It means accept, be patient with, be tolerant, trust. It means *give it a chance.* It means *hear somebody out. Let him make his point.* It can mean *try it.*

There's a new kid in school. Openness is making the first move to meet him.

You're in a Chinese restaurant looking at an elaborate menu. You recognize "chop suey" and "chow mein" but you order "shark's fin soup with quail eggs" and "full moon dumplings." That's openness.

There are two kinds of openness in dialog. The *speaker* has one kind of openness, the *listener* has another.

The speaker is open if he fairly evaluates the importance, truth and pertinence of his information and if he fairly estimates his ability to deliver it. An open speaker respects his listener.

The listener's openness is characterized by attention, interest and sincere questioning.

When both speaker and listener are open, they are in *I-Thou* dialog.

Open to what? To *differences* and *change.*

1. By *difference* I mean opposition, denial, rebellion, contradiction. Difference is *rejection*, either partial or total rejection, either temporary or seemingly permanent rejection.

In dialog, I try to be open to a listener who disagrees with my opinions, whose interests are foreign to mine, who smells different from anyone I like, who is different in color and looks, whose attitude contradicts mine, whose power conflicts with mine, whose manners annoy me, who competes with me in talent, who comes from a different background and neighborhood, who was educated by different standards, who argues by different rules, who prays to God in a different way with a different accent, or who doesn't pray and ridicules my prayer.

The more open I can be toward a listener's differences and changes, the more can I communicate and the more I *become.*

My becoming in others depends on my acceptance there. I may be totally right, but I can be rejected anyway for everyone is free to reject right and choose wrong.

I cannot force others to listen. I can only attract them to listen.

2. *Change* is also a kind of difference but it does not necessarily imply rejection.

In dialog, I try to accept change in someone who used to look a certain way I got used to but who now looks different; who once was my friend but has now forgotten me; who used to telephone regularly but has stopped; who once liked a certain music or action or place or food but who now doesn't; who was once alive but is now dead, was once young but is now old; who once loved me but now doesn't.

Change includes change in myself. I had a wonderful tan a month ago but I've lost it and I look like oatmeal now. I felt very religious last year but not this year. Last year I couldn't wait to finish school and go to college. This year, I don't know. I feel more like getting an airline job and traveling. Last year I liked this girl till I thought I'd go wild. This year I can't stand the sight of her.

What can keep me from being open? False images, misinformation, prejudice, bias, clichés.

Openness is controlled by images. *Images are the valves and filters of dialog.* A narrow, incomplete, thoughtless or illogical image closes the speaker or listener. It brings along feelings which also have a closing effect: discouragement, cynicism, sarcasm, nit-pickyness, suspicion, vengefulness, doubt and nervousness.

A prejudice, to take the word apart, means a *prejudgment.* It is a judgment made in advance of an actual experience or ahead of actual evidence. The pre-judgment might never have been made by the prejudiced person but simply copied from parents.

You've got to be taught
To hate and fear.
You've got to be taught
From year to year . . .
You've got to be taught before it's too late
Before you are six or seven or eight
To hate all the people your relatives hate,
You've got to be carefully taught.

Pre-judgments are necessary for human action. If we had to put our hands in the gas burner each time before judging that burning gas is hot, we'd soon be without hands while still lacking wisdom.

Pre-judgments make trouble if they are misapplied, though even then they can be corrected. It is when a prejudice is grooved in deep and snug and is associated with strong feelings that it harms us.

Pre-judgments harm us in two ways: they distort both decision-making and communication.

A bias favoring someone may be just as bad as a prejudice against him if it blinds us to seeing him as he really is.

Experiments by the late Boston psychologist, Dr. Gordon Allport, show that deeply prejudiced persons are slowest to detect differences and equally slow to accept change. One of Dr. Allport's tests for example, employed a box of fifty picture cards. The first card shows a dog chasing a cat.

"What do you see?" Dr. Allport asked his subject. The subject handled the card, examined the picture, then said: "A dog chasing a cat."

Next card: "And now?"

"A dog chasing a cat down the sidewalk."

Next card: "Now?"

"A dog chasing a cat up a tree."

A dog continues to chase the cat for twenty or so cards. But now look carefully. There is a slight change in the drawing. The dog has the cat's ears and the cat has the dog's.

Three cards later the tails are exchanged, then the paws one by one. Then the markings on the body and finally the body itself. The *two animals have switched positions.*

As he showed each new card, Dr. Allport would ask "What do you see?" The subject continued to answer: "A dog chasing a cat over a log . . . into a hole . . . into a store. . ."

The last five cards in the set show a cat chasing a dog, but the test person continued to "see" the dog chasing the cat.

How do prejudices affect communication?

1. Prejudices clog our impressions. They foul our expression. They choke and stunt *becoming.*
2. Prejudices attack openness. They are blocks. They stiffen our attitudes so that we don't quickly respond to difference and change.
3. They affect the listener's openness too. A prejudiced listener in turn affects persons trying to communicate with him.
4. Prejudices attach themselves to images, both the self-image and our images of others. Prejudices thus make dialog impossible since they interfere with both the *I* and the *Thou.*

PREJUDICES: TWO VIEWS

Without the aid of prejudice and custom, I should not be able to find my way across the room.

William Hazlitt

We see only what we are ready to see, what we have been taught to see. We eliminate and ignore everything that is not a part of our prejudices. J. M. Charcot

---

Dialog vs. Communication

What is dialog and how does it differ from communication?

Communication relies strongly on the senses since it uses signs and symbols, gestures, winks and false eyelashes, clothing, perfume, badges, pins and rings. Touch plays an important role in communication. Touch, gestures and symbols may stimulate dialog to start but as steady vehicles they are awkward. Spoken face to face or written, *words* work best.

Communication can be vague and inexact. Music and art, interior decoration, landscaping may all communicate some feelings but they are too clumsy for dialog. *Dialog is precise communication.*

A person can communicate with strangers in a foreign country though he doesn't know the language. He can signal that he wants water or food or gasoline or a place to sleep or a garage. But he has to know language in order to dialog, and the better he handles language the better will he be able to dialog.

Dialog demands that he be able to represent his Actual and Potential SELF. The more exactly he knows words, the more exactly will he be able to describe his SELF. The richer his vocabulary, the better his chance of being able to address others.

Dialog searches out *interior* and *essential* contact. Communication may be satisfied by a meeting of Actual SELF and Actual SELF. Dialog attempts fuller, deeper, meetings which enrich the Potential SELF.

Animals communicate with each other and with man. They do not dialog. Only man can dialog.

I can talk my brains away without even noticing that you're not listening. I can talk about myself without knowing who I really am. You and I can *converse* about sex without honestly daring to say what's truly on our minds.

Dialog is not blabber. It's different from mere talk, more than conversation. It demands attention of both listener and speaker.

The greater part of what is today called conversation among men would be more properly and precisely described as speechifying. In general, such people do not really speak to one another, but each, although turned to the other, really speaks to a fictitious court of appeal whose life consists of nothing but listening to him.

Martin Buber

Talking a lot with nothing to say is worse than having something to say and not saying it.

Boy, 18, Milwaukee

Constant talking gets pretty sickening.

Girl, 17, New York

Dialog does not mean saying *everything* that's on the mind. Many thoughts which waft up from the subconscious or slip through the mind are immature, uncharitable, even vicious. It's not necessarily honest or wise to blurt them out just because they're there. They belong to the SELF as fragment-judgments or as reject-decisions.

Inter-human truth does not depend on one saying to the other everything that occurs to him, but only on his letting no *seeming* creep in between himself and the other.

Martin Buber

Because communication is vague it can be frustrated by *ignorance* and *misunderstanding.* It can be misled by prejudice and bias. It is easily discouraged by *rejection.* Dialog by definition seeks to avoid these three sources of failure. Symbols are ambiguous. We can be ignorant of their meaning or we can take a wrong meaning from them. Dialog tries to make sure that the symbols used are clear, not only to the one who uses them but to the one they are intended for.

Dialog is deliberate. Contact in dialog is intended. Communication's contact is often accidental. Radio and television broadcasting, book, magazine and newspaper publishing are all called "communications" but they are a one-way scattering of expression. The senders hope for understanding, sympathetic receivers.

Communication does not exclude honesty but it doesn't demand it as dialog does. Dialog is talk which *engages.* It puts *me* on the line and meets *you* there. *I* level and expect *you* to level too.

I can communicate false information or a false image of myself but dialog by definition forbids both of these. I can encourage you to go on talking about yourself without even letting you know I don't believe a word of what you say. I can let you talk for hours about your car or your family without telling you that you bore me.

The boy described below fakes an interest in the girl's wishes.

On one date the boy did nothing but ask me what I wanted to do and we couldn't come to an agreement because he just did what

VLADIMIR

I liked. But he wasn't interested in what I liked. And I wasn't interested in what he wanted to do when I finally found out what it was. It was our first and last date.

Girl, 18, St. Louis

Communication between male and female persons may be exterior and superficial: a flash of symbols, false impressions, false images, bravado, flattery, artificial talk and laughter. But dialog searches out *interior* and *essential contact.*

Man is willing to accept woman as an equal, as a man in skirts, as an angel, a devil, a baby-face, a machine, an instrument, a bosom, a womb, a pair of legs, a servant, an encyclopedia, an ideal or an obscenity; the one thing he won't accept her as, is a human being, a real human being of the female sex.

D. H. Lawrence

A bad date is a date which consists of only dealing with sex just for the fun of it not really liking that person and trying to like him . . .

Girl, 17, Milwaukee

A bad date is one that shows no interest in you as a person and just wants to see what he can get.

Girl, 17, New York

He treats you like one of the guys instead of someone he asked out.

Girl, 18, St. Louis

The boy thought of the girl just as a girl (body) and not as a person. The boy should respect you for being a girl, a person not just somebody in a skirt and sweater.

Girl, 18, St. Louis

He isn't interested in you as a person but just as a member of the opposite sex.

Girl, 18, New York

I-Thou meetings, Buber said, are "strange, lyric and dramatic episodes, seductive and magical." But *dialog is work.* It sometimes takes nerve.

Failure happens when I can talk to a boy about everything but ourselves and what we mean or don't mean to each other. And how right or wrong our sex relations are.

Girl, 18, New York

. . . both of us have to give of each other to make the date come out right. Communication is everybody's job.

Boy, 17, Denver

Conversation should flow and if it doesn't somebody isn't making an effort. I go up the wall when the other party is stiff and doesn't loosen up.

Boy, 17, Milwaukee

I feel uncomfortable where there is dead silence. There is obviously not too much exciting going on between the two. They should talk to find out what is wrong. They have to try.

Girl, 18, St. Louis

I dated a guy who was very intelligent and active in school. He didn't ask me out again. I think because although I was interested in him, *I* didn't let him know I was interested.

Girl, 18, New York

My date and I had a misunderstanding and for a long time we both didn't talk either because of our pride or just stubbornness. If we would have talked it out right away I think we would have had a most enjoyable evening.

Girl, 17, Denver

I went out with this boy a couple of times and one night, while we were still on the date, he asked me if I could go out again next week. I told him I had already made other arrangements. He was so mad that he didn't talk the rest of the date. And I didn't find out until the next day what was his problem. If he would have talked it over or said he was displeased maybe some of his anger could have been let off.

Girl, 18, Milwaukee

Barriers to Dialog

What keeps the *I* from meeting the *Thou*? What are the barriers blocking dialog?

1. *Disinterest: Apathy*

Dialog is work. You have to be interested enough in the other person to *want* to dialog with him.

2. Excessive *competitive* spirit. Some persons value victory over the other person higher than they value dialog. They regard others as potential *Its*.

Dialog isn't argument or debate because the *I* does not try to win over the *Thou*. Dialog is more interested in truth than in winning an argument. It is honesty without uncharitable rudeness. The dialogist isn't chagrined at being proven wrong. At least he's learned what's true.

I dated one girl who cut down almost everything. My friends, their dates, cars, she didn't want to do anything we did and acted superior and demanding. These last two I think I hate most in girls. Superiority and demanding qualities.

Boy, 18, Denver

3. Lack of language ability, as in a foreign land. Poor vocabulary. Language inability also affects the self-image.

4. Physical fatigue which distorts self-image and other-image.

5. Alcohol and drugs which distort images.

6. Prejudice. Bias. Clichés.

7. False, faked or masked images whether of Self, or Other, set up the most serious blocks against dialog. The image stimulates fear, distrust, suspicion and dislike.

WHAT IS THE BEST WAY TO CALL OFF A DATE?

How to cancel a casual date usually is a little decision involving communication. Like all little decisions it gets the person ready to make serious decisions.

How do you call off a date—by telephone or in person?

Use the telephone if there's an emergency. Don't use the phone or the mailman to avoid a painful or embarrassing meeting.

> I believe in breaking dates by talking to the person if possible and telling him what you have to. A phone call isn't very good because the person can't see if you are really sorry you have to break the date or not.
>
> *Girl, 17, St. Louis*

> The best way to call off a date is merely to phone and tell her the good excuse you have. If it is a big date (something important) you should meet her and tell her you have to break it.
>
> *Boy, 18, Denver*

> I'd never call off a date, unless it is an absolute must. I would talk to the girl rather than call her on the phone.
>
> *Boy, 17, St. Louis*

It takes both experience and ability to say certain things face to face. Persons who have bad communication images often find it easier to give compliments, to show love or gratitude, to reveal deep thoughts and feelings by letter or by telephoning rather than in person. They may be fooling themselves, however, if they believe they are in dialog.

> Dan is a swell guy, but he really is dead when it comes to conversation. While Dan is away at college, we write letters. His letters are terrific and interesting but when he's home and we go out it's like dating two different people. I try to strike up a conversation that he could enjoy and really talk on but he answers in the shortest form. It's hard to have fun when there is no conversation. Our main communication is talking so if one can't communicate it isn't too good.
>
> *Girl, 18, New York*

> Boys don't like girls who lie or stand them up—a little honesty can help your reputation a lot.
>
> *Girl, 17, Milwaukee*

Don't accept the date to begin with if there are doubts about keeping it. If an emergency arises or you find out the party is married or going steady or of bad character, be blunt and tell him exactly why you are refusing this date.

Girl, 18, St. Louis

Level with the guy. If you respect him you will be courteous enough to give him the truth so he won't hear it from someone else.

Girl, 18, St. Louis

Don't make the guy feel it's the "brush off" always consider his feelings . . . don't be mean or sarcastic.

Girl, 17, Denver

Call her a day or two before hand, tell the truth because if it were a lie she might find out, then you're *through.*

Boy, 18, New York

Tell the real reason. Even if this be the awakening to the realization that you do not like the girl.

Boy, 17, Milwaukee

Because you would rather date Sue, don't break a date with Mary. If you keep this rule, you won't have to worry about what to say.

Boy, 17, New York

Breaking the Steady Date: A Crisis of Dialog

How do we break bad news? How do we take it when somebody brings it to us?

During marriage, there are many times a husband and wife will have bad news to exchange.

BAD NEWS	REACTION
"I've been fired!"	Run out of the house and get drunk.
"I had an accident."	Sit in front of television and get drunk.
"The baby ate a bottle of aspirins. I left her alone to go to the store.	Call up mother and talk with her for a couple of hours.
When I got back she was lying on the bathroom floor in a coma."	Scream.
"Mother is coming to stay with us."	Break a dish.
"We have to move to another city."	Blame each other and fight.
"I must have miscounted—I'm pregnant again."	Shrug shoulders and go back to reading the paper.

As little children we have seen our own parents bringing home bad news. We'll never forget some of the scenes that followed. They influenced our own communication and may still influence how we break painful news and how we react to it.

WHAT IS THE BEST WAY TO BREAK OFF A STEADY DATE?

Pulling out of a steady-dating pattern is much more difficult than merely calling off a casual date. As a crisis, it tests both decision making and dialog.

All of us are aware that rejection can hurt cruelly. Some persons have a neurotic wish to cause such hurts, but we don't normally want to inflict this agony on someone else. At the same time, we like to be truthful. We don't like to lie.

> We felt ill at ease with each other because we knew we were going to break up because we weren't having that much fun with each other anymore. We were just getting sick and tired of each other. After we broke down and talked about breaking up it cleared the air, but the date was still ruined.
>
> *Boy, 18, Denver*

> The only way is to level with a boy all the way.
>
> *Girl, 18, New York*

> If the girl tells him why she broke it off, the next time he goes with someone he will know what to do and what not to do. He will learn from her what he is like to other girls.
>
> *Girl, 18, New York*

> I think that if two persons know each other very well, they can tell each other anything including their feelings for each other. If she's stopped liking him she will tell him and vice-versa.
>
> *Girl, 17, St. Louis*

> It's hard for a girl to come out with it and tell the boy she wants to date others. She is afraid of what he'll say, do, or how he will act toward her.
>
> *Girl, 17, Milwaukee*

> Be honest you owe that much to him.
>
> *Girl, 18, New York*

> If it's for financial or family matters, tell the girl that. If it is because she treats you like a crud and talks behind your back, tell her where to go.
>
> *Boy, 17, Denver*

> Call the person up and explain your point. Or if you are too chicken then write a letter. This is a silly question because if boys and girls were taught sex education there would be no steady dating at such an early age.
>
> *Boy, 17, Denver*

> Use the telephone to break off a steady date. It's such an agony to see somebody hurting. You might give in.
>
> *Girl, 18, New York*

Use the phone. If you do it in person the chances are you won't break up but will keep going steady out of pity.

Boy, 18, St. Louis

. . . Go and see the person and explain. I would rather have a person come and tell me instead of just calling.

Girl, 17, St. Louis

The best way I can think of is to have a serious heart-to-heart discussion. It has to be in person. Don't write a letter or anything. Say you're sorry and be as honest as is safe to be. If the couple cares for each other, and respects each other as much as their dating patterns suggest, they will listen and respect each others feelings.

Girl, 18, New York

I honestly don't know if there is a best way to break off a steady. This is my present problem and I wish I knew the answer. There is no best way, without a nice fight. You just break it and of course she feels entitled to know why and even if you have a truthful reason she won't believe it so she questions you and you fight.

Boy, 17, St. Louis

See less of her and try to pass her on to a friend.

Boy, 18, Milwaukee

You should go to a quiet place, where outsiders will not disturb you. You should then explain that you don't wish to continue seeing her and give her a reason why. If I was still interested in the girl, I would try to set up a new date for her with somebody else.

Boy, 18, St. Louis

Tell her you don't want to go with her anymore and that you found someone else and you don't like her anymore. Tell her the truth.

Boy, 17, Milwaukee

Just go ahead and date someone else. Tell the other person you don't want to go steady and that's that!

Boy, 18, New York

Ask for your ring or pin back. If you really want to do it, do it quickly, if things are fairly bad. Explain and if he or she doesn't like it—tough. Telling the truth might be terrible but to some it is best in order to avoid later complications. Come out with your reasons why. Don't keep stalling. Tell her you don't think it would work and you want your ring back.

Boy, 18, Denver.

Fifteen Dialog Dodges

Dialog takes nerve. You may be too chicken or too lazy to meet its demands. There are many ways to get around it. Here are fifteen tested, safe and easy dialog dodges:

D.D. #1 KEEP BUSY

I used to take out this one girl frequently to nice places and usually gave her a good time. But on our last date she acted strange—we were on a boat ride and then went to roast some hot dogs. And that is all she did all night was roast hot dogs, even though other people offered to help. I was left standing in the cold.

Boy, 18, Denver

D.D. #2 BLAME SOMEONE ELSE— YOUR PARENTS FOR EXAMPLE

Tell him you want to go out with other guys, not with just one person all the time. Or tell him your parents don't approve of you going out with one guy (most parents do feel that way).

Girl, 17, St. Louis

I'd say I still like him very much, but for one reason or another (I want to date others; my parents objected too much; it is not good for us to be together so much) I'd have to break it off but I would still like to go out with him.

Girl, 18, Milwaukee

D.D. #3 BE VAGUE

Explain how your ideas have changed and that's why you think you should break up.

Girl, 17, St. Louis

If you don't like a boy, give hints to the affect that maybe next weekend you'll be busy and won't be able to see him. After awhile, he'll get the message, loud and clear.

Girl, 18, New York

D.D. #4 RELY ON THE OTHER PERSON TO "UNDERSTAND" WHAT YOU'RE AFRAID TO SAY

This girl never *said* she wanted out. I was supposed to guess it. She just acted heavy and bored. I called her less and less, then stopped completely. I got her message but I hated her way of delivering it.

Boy, 18, St. Louis

D.D. #5 NECK

Communication was stopped completely after the first two hours because I was fighting to keep him off me and after I did get him off—I moved over against the door and kept my mouth shut—he didn't say a word.

Girl, 17, New York

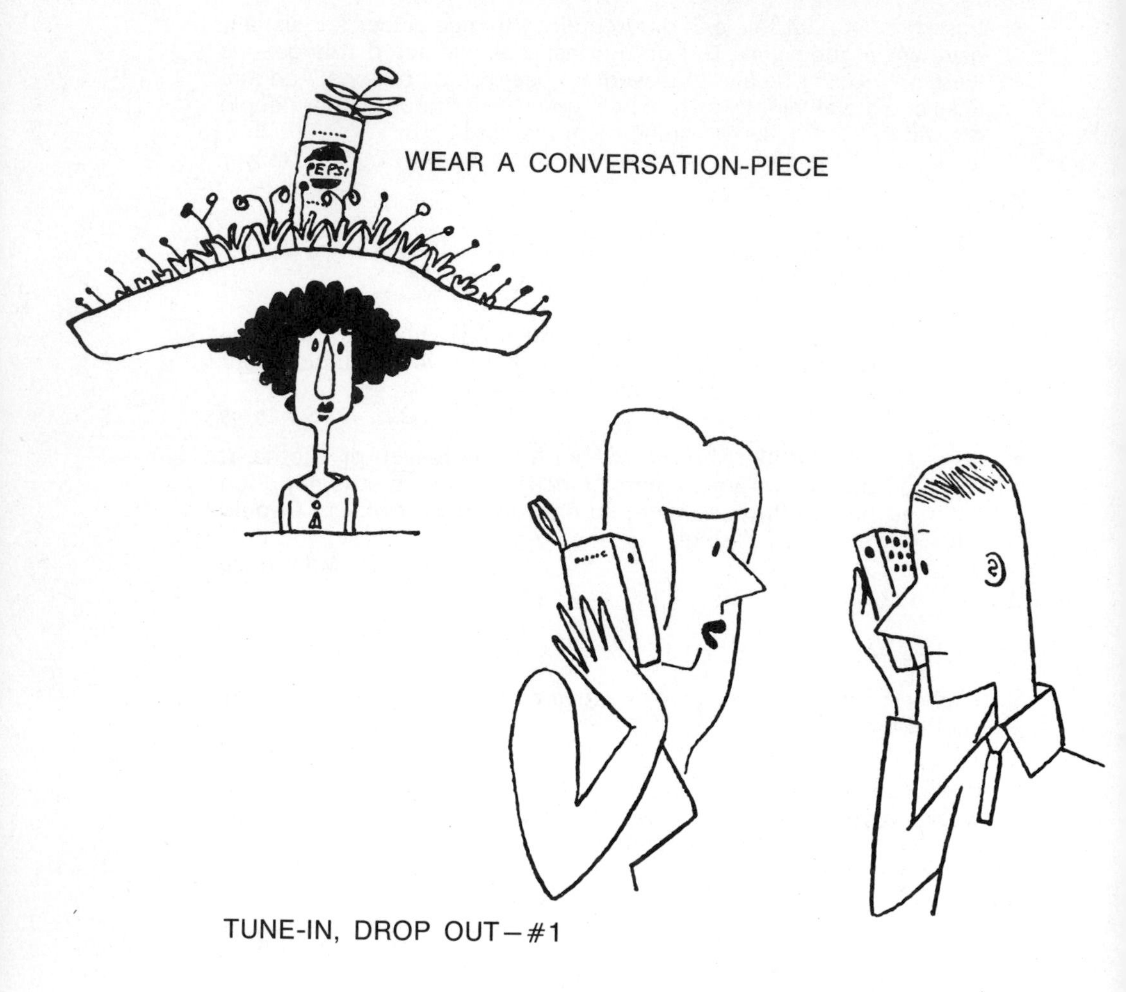

WEAR A CONVERSATION-PIECE

TUNE-IN, DROP OUT—#1

SEE A DRIVE-IN MOVIE

WEAR A MASK

TUNE-IN, DROP OUT—#2

BE BOSS TALKER

VLADIMIR

Either the guy doesn't talk, or he talks too much (brags about himself too much) or else he hangs all over you all night—wants to neck, etc.

Girl, 17, St. Louis

. . . is boring, cannot speak with you or is giddy and emotional and finally likes to play kissy-face all night.

Girl, 17, Milwaukee

I remember one time—the guy was completely disinterested in everything I'd mention, including sports, draft, etc. All he wanted to do was neck. The entire relationship was a failure.

Girl, 18, St. Louis

You have to do all the talking and the only thing he wants to do is neck. He only talks about himself and also who wants to go only to the park for a date. If a boy is like this, he sure isn't going to be popular because word gets around. No one will go out with him.

Girl, 17, Denver

D.D. #6 TALK NON-STOP

I dated a guy who was very good looking, had an offbeat though good sense of humor but he talked non-stop! He wouldn't shut up all night. It wouldn't have been so bad but he talked incessantly of music; popular and rock 'n roll. I couldn't understand a word he said because he talked about instruments and singers. And he never, never stopped.

Girl, 18, New York

You get mad at your girl because you are not talkative and can't stand her talking your ear off.

Boy, 17, New York

D.D. #7 KEEP THE TALK ON YOUR SIDE OF THE COURT, IN YOUR POSSESSION

She talks about herself and hardly listens. Criticizes too much, generally keeps the other person out of the whole scene. That's a bad date.

Boy, 18, Milwaukee

Your date wants to do all the talking. He will ignore what *you* say and will say what *he* wants . . . your date constantly talks about himself and doesn't let you say a word.

Girl, 17, St. Louis

All he could talk about was school. Everytime I tried to bring up another subject he would go right back to his school. I had the worst time I have ever had. I couldn't wait to get home.

Girl, 18, Denver

D.D. #8 TALK FACTS, NOT MEANINGS

My date and the other couple talked amongst themselves about cars which I knew nothing about. The date was a flop.

Girl, 17, Denver

D.D. #9 WAIT FOR THE OTHER PERSON TO START

There is little conversation and both parties are unwilling to get the ball rolling.

Girl, 18, Milwaukee

D.D. #10 USE BIG WORDS. USE WORDS ONLY YOU UNDERSTAND

The boy was four years older than myself. He was a junior in college and he used all the real big words which I did not understand. And I had to guess at what he meant.

Girl, 18, Denver

D.D. #11 THROW A SUDDEN MOOD

I once took a girl out to eat. We were talking and laughing on the way. When we got to the restaurant and sat down she clammed up. I didn't know what to do. When I asked her a question she'd answer it as short as she could make it, and then be quiet again. I was never so unhappy.

Boy, 18, St. Louis

D.D. #12 BE EASILY HURT AND SULK

One person hurts the other unintentionally. Then when you ask: "What's wrong?" the hurt party says "nothing." Then there is no more conversation and it's the end of the date.

Boy, 18, Milwaukee

D.D. #13 BE TIRED

She wasn't in a mood to talk. Every time I'd say let's do something or go, she'd say, "I'm too tired." She didn't talk. She was too tired to smile. I got disgusted and brought her home very early.

Boy, 18, St. Louis

D.D. #14 LAUGH IT OFF

He took everything I said as a gag. Ha Ha. All the time. I could have killed him. He never got serious except necking.

Girl, 17, Denver

D.D. #15 DON'T LISTEN

I had the feeling she wasn't with me. Her eyes were off somewhere. I'd say something, she'd just answer "umm, that's right." I tested her once. I said "Do you think Jim is pregnant? He looks it." She said, "umm. That's right."

Some Rules for Dialog

1. Don't avoid dialog. Dialog is hard to start and hard to keep up; but if you miss it, you miss a chance to *become*.

2. Trust the other person to level with you. First trust that he's trying to level with himself. Believe he means what he says and that he believes your meanings.

3. Dialog relies on language—not touch, not gestures, not facial expression, not intuition, not music, not art, not symbols—words talked face-to-face. Don't *assume* you're communicating. Don't assume everyone's listening to you.

4. Observe rules of logic. If you generalize, be extremely careful not to over-stretch your generalization. Be ready to modify your generalization if evidence suggests you should.

5. Respect your word. Don't give it easily. Don't promise what you can't deliver.

6. Respect your name. Don't sign it carelessly. In modern organizations we use numbers to speed the finding of information. We have zip codes, social security numbers, license plates, bank account numbers, registration numbers. Don't let your SELF become a number. Don't let your SELF be addressed as a number. A number is an *It*.

7. Don't treat the other person as an It. A Thing. A Category. A Class. A Sex. A Political Party. A Religious Sect. A Uniform. Look at each person as a true human being with fears and loves and hates and angers, with goals dearly cherished and desired, with ambitions, with special and important needs.

> The chief presupposition for the rise of genuine dialogue is that each should regard his partner as the very one he is. I become aware of him, aware that he is different, essentially different from myself, in the definite, unique way which is peculiar to him, and I accept whom I thus see, so that in full earnestness I can direct what I say to him as the person he is.
>
> Martin Buber

8. Don't hide your Self in a category. I am a boy. I am a girl. I am a Catholic, Protestant, Jew. I don't believe in any god. I believe in One God, the Father Almighty. I am a Republican or Democrat, a Conservative or Liberal. But at the same time I am an individual human being, educated by my unique mother and father in our unique family. I am me. I am myself, different from you and from everyone else.

Speak for your Self, not for a false image of yourself nor for a false image you believe the other person may have of you nor one you want him to have of you.

> Rabbi Zusya said a short while before his death:
> "In the world to come I shall not be asked:
> 'Why were you not Moses?'
> I shall be asked:
> 'Why were you not Zusya?' "
>
> Martin Buber.

---

9. Listen to the person, not his appearances. Check your prejudices and biases. Keep in mind some persons may want to communicate to you but don't have the skill or vocabulary to be able to do so.

10. Don't *impose* or force your way of thinking on the other person. You may have authority, power, strength and position. You may be smarter, more popular. You may be right. But if you force your attitude on others you don't respect the other person's right to be himself.

11. Address the person's Self and not your own false image of him, nor his own false image of himself, nor the image others have of him.

Here is a classmate, a "C student" according to his report card. When you speak to him, do you address him as having "C" ideas? Do you expect others to accept your ideas as "A" ideas merely because you're an "A" student?

> To be aware of a man, therefore, means in particular to perceive his wholeness as a person determined by spirit; it means to perceive the dynamic center which stamps his every utterance, action and attitude with the recognizable sign of uniqueness.
>
> Martin Buber

THE DIALOGIC PERSON

Dialog is an exchange of complex decisions. Step by step, I put choices in front of you. Yes or no? Do you agree or don't you? Do you understand or don't you? What do *you* think (now that I've told you what *I* think)?

Matching me step by step you assert your stand and put the same choices to me. If I can take these steps, I grow because I *become*. If you take the steps, you *become*. I help you by making the steps honestly lead to me.

Dialog is a *double becoming:* (1) The Actual Self of each communicating person is engaged; (2) The Potential Self of each becomes *Actual.*

What is the product of dialog? Who is the dialogic person?

* The dialogic person is a total, authentic, committed person.

* He doesn't impose his ways on others. He states his views respecting the differences of others.

* The dialogic person is an open person, able to grow and develop more and more. He hungers to engage himself with life and with people, believing that Self knowledge and Self *becoming* are possible only through such engagements.

The *I* becomes a *Self* only in relation to a *Thou.*
Martin Buber

---

*The dialogic person distinguishes his actual Self from his Potential Self.

*He sees his Self in perspective, which means he has a sense of humor.

*His words mean what they say. If he uses clichés, he uses them with thought.

*He is able to promise.

*He is able to forgive wrongs against him.

*He recognizes his Potential Self has unexplored untried powers. He knows he has untested weaknesses and is prepared to discover them by his own inquiry or through the insights of others. He is a mature person able to assume responsibilities.

* He regards others in the same way, distinguishing who they are from what they appear to be, who they actually are but being open toward recognizing the persons they could be.

He sees each of these individuals as in a position to become a unique, single person, and thus the bearer of a special task of existence which can be fulfilled through him and through him alone. . . . He is a helper of the actualizing forces.
Martin Buber

---

* He is able to face and endure the truth wherever and in whomever it shows itself, believing that truth leads to the Face of God.

This is the ultimate purpose: to let God in. But we can let him in only where we really stand, where we live, where we live a true life. If we maintain holy intercourse with the little world entrusted to us, if we help the holy spiritual substance to accomplish itself in that section of Creation in which we are living, then we are establishing, in our place, a dwelling for the divine Presence.

Martin Buber

Touch Communication

A person's ability to communicate grows as the person himself grows. His first experience with communication is in the womb through *touch*, not with his fingertips, but with his whole body inside and out, through the sensation of warmth, through his mother's pulse, through her muscles when she tightens or relaxes them in the rhythm of her activities. *Touch communication is tightly tied to his first Basic Sense of Good Feeling.*

Hunger, cold, and rejection from the womb, collision, smothering, restriction, these are the shocks that assault the newborn infant. He cries and his mother answers by holding him. He stiffens his body and his mother is warned that he's afraid or doesn't like something. He feels his mother's gently throbbing warmth against his skin and "remembers" dimly how secure his womb life used to be.

In time he gets used to being out of the womb and replaces his old basic sense of good feeling with a new one. Taking in food is made up of both old and new experiences. The *old* familiar heartbeat, rhythmic rocking and warmth; *new* voice sounds, pressing, kissing and stroking.

The infant makes an important discovery—that he can get the body-to-body feeling he wants by *reaching*. At first his reachings are aimless and touch is accidental. But, after thrashing about, swinging his arms and legs in all directions, through clutching and grabbing, he gradually learns how to *aim*. Now he can search for nearness. He doesn't have to lie helpless waiting for someone to find him.

The infant also learns how to *keep* this nearness. He learns to cling to his mother, to grasp and to hug. By means of *reaching* and *holding*, he can get back some of that reliable warm cadence which dominated all his early life, his mother's heartbeat.

Again and again, the baby will reach toward someone or some object far from his grasp. As he learns to know his reach, he learns he can get what he wants merely by *pointing*. Pointing is a short-cut way of touching.

One day the baby discovers he can wrinkle his face as if he's going to cry or laugh and get results without having to go through the fuss of crying. This, like the discovery of pointing and gesturing, is an important discovery. The baby finds he can communicate by *facial expression*.

Words, gestures and facial expressions save time, he discovers. Before he could manage them, he had to wait till his mother guessed what he wanted. Now he has better control over his life. He gets better results and gets them more immediately.

Every family has certain facial expressions, words and gestures which have meanings only to family members. Every neighborhood, too, has its special language of signs and sounds unique to the neighborhood. As the child grows, he will have to learn new languages and retire old ones; first his family's, then his neighborhood's, then his grade school's.

All through life touch is an important basic form of communication. It is man's medium of communicating with most animals.

Ivan Tors, a Hollywood producer of animal films (Africa —Texas Style, Flipper and others) is famous for pairing a python and a chimpanzee, a lion and elephant, as well as a tiger and fawn in the same cage.

Tors believes in what he calls "affection training" as opposed to the old whip-and-fear methods of handling animals. His training system begins in the nursery where his attendants spend the day stroking and fondling young animals. The more dangerous species of animals are stroked with long, sponge-tipped "petting sticks." These sticks are reduced in length until the attendant can tickle the animal with his fingers. Tors tells the attendants: "You cannot love without touching."

Touch can relax tensions and dissolve arguments. It can also create tension and cause arguments. Some persons can communicate merely by *being present* to one another.

> If two people are alone and they really like each other there doesn't have to be communication between them. Just being together is all that really matters to some people. Some people can get awfully tiresome when all you do is hear them talk.
>
> *Girl, 18, Denver*

But communication limited to touch is unsatisfying and disappointing. Compared with language, touch is as vague as an infant's grab for warmth and nearness and care. Touch may satisfy the body's basic sense of good feeling, but the Self is left unexpressed.

For this reason the excitement of necking-dates is shallow, short-lived and ultimately disappointing.

> I can't stand it whan a guy forces himself on me. If a guy wants sex and nothing else and that's all he took me out for, I want to go home.
>
> *Girl, 17, Denver*

> We go out but we're just two bodies. Never talk, only neck.
>
> *Boy, 18, Milwaukee*

The Self has memories to release, plans to discuss. It has dreams, worries, fears, and hopes that can't be easily expressed by glances. Holding hands with someone can be a simple, wonderful joy but it won't do as a substitute for language if you want to talk about the draft or college. A caress says a lot but its meanings are limited. The stroke of fingertips over skin is a poor substitute for words. The kiss is not the best vehicle for telling a joke.

Loneliness, the feeling of being different and isolated, can sometimes be assuaged and sometimes forgotten by body-to-body nearness. But unless there is Self-to-Self nearness loneliness soon creeps back. The grand medium for human intercourse is undeniably, inevitably and necessarily *language*.

The Sex-Centric Date

Girls frequently complain that boys have nothing but sex on their mind.

> A bad date is centered around sex. There is nothing else to do because one or other party is a bore so you turn to a common factor—sex. This is not the purpose of a date. A good movie or conversation is a good date. Sex need not be eliminated but it should not rule.
>
> *Girl, 18, New York*

> To me a bad date is going to a movie and not seeing any of it.
>
> *Girl, 17, Milwaukee*

According to one survey of opinions about dating, more girls than boys preferred double dates over single dates. Girls say they double date to protect themselves from bores and "sex maniacs."

He's very shy or he's overly aggressive. Either way all he wants is heavy necking or petting.

Girl, 18, St. Louis

	Boys	Girls
More fun double dating	53	74
More fun single dating	43	22

Purdue Opinion Panel

I can relax on a good date and enjoy myself, feel at ease. One where I do not feel uncomfortable about silence or don't have to worry about screaming for help. I like to go out with someone I've known ahead of time.

Girl, 18, New York

Going out with someone who has one thing on his mind, sex, and having him try to make passes all night. You have to watch him every minute and can't relax.

Girl, 17, Milwaukee

I avoid boys who play tough guy, a Romeo or a sissy. A boy who doesn't know and expect limits in anything.

Girl, 18, St. Louis

Watch out for one who's too aggressive and paws all over you . . . takes you out to see how far he can go with you.

Girl, 18, New York

Kissing, nuzzling, stroking and squeezing, when these actions stimulate specially sensitive areas of the body, are all forms of sex play. *Necking* is light and affectionate sex play. *Heavy necking* is intense and prolonged sex play. Both stop short of actual intercourse, though "heavy petting" may result in orgasm. The function of such stimulation is to prepare the body for sexual intercourse, though intercourse might not be specifically intended.

Girls rarely force themselves on a boy in the same way that a boy may force himself on a girl. They will rarely have to. Out of a thousand statements written by boys about girls, only one or two complained of a girl who had sex on her mind or was "all hands."

I was lying in the car in the drive-in and she didn't want to stop, and I wanted to talk and she wanted to keep going, so that's a lack of communication.

Boy, 18, Denver

The boy's usual complaint is that the girl is cold or aloof and won't concede even the smallest sign of affection.

. . . you can't get along with her and constantly fight, or you take her somewhere that she doesn't like. And when you take her home she quickly runs to the door without a good-night kiss.

Boy, 18, Milwaukee

The girl was immature or inconsiderate or acted too much like a virgin. Not being able to hold hands in public . . . the girl is a prude.

Boy, 17, Milwaukee

Your date sits at the other side of the car all night and you take her right home after a show, etc. . . .

Boy, 18, Denver

. . . the girl doesn't talk and you don't at least get a good night kiss (first date) I like it when the girl is very friendly and you get some good lovin'.

Boy, 18, New York

When the girl complains of having to fight off a boy all evening, what is often implicit in her complaint is that she didn't like the boy anyway. If a girl likes a boy, she doesn't object to being kissed. Boys know this. What they're not always sure of is whether the girl likes them or not. What does a girl's display of friendliness mean? Is she encouraging him to be bolder? Some girls will tease a boy forward, then cruelly slap him back into place. Boys get wary.

Not all boys are sex-centric or self-centric either. Papers written by boys show most of them have simple tastes and enjoy a casual good time.

I can't see how a person can have a bad date. You take a person out if you like her and if you like her you're bound to have fun.

Boy, 17, New York

When I get home and can honestly feel that I have had a good time and gave the other person a good time, that's what I like, lots of laughs.

Boy, 18, Milwaukee

I blame myself for bad dates. It's a bad date when the girl has not enjoyed herself because of my fault.

Boy, 18, St. Louis

Your date has to have a good time and so do you.

Boy, 17, New York

Having a lot of fun together, even if it is just talking and sitting at home.

Boy, 17, New York

A date is a give and take for both parties. Try to make your date feel at ease and have some enjoyment . . . not only by just being together but also the place you go should be proper and fitting.

Boy, 17, Milwaukee

A bad date is when a girl refuses to relax in my company and instead tightens her grip on her actions by being timid, shy, that sort of thing. I don't say she has to go to the extreme of sexual relations.

Boy, 18, St. Louis

A good date is just go out and enjoy yourself not one where you have to necessarily be "madly in love" with the girl.

Boy, 17, Denver

Keep Out of Steamy Dark Corners

Some girls pride themselves in being able to turn off their passion in time. Whether they can or not, the practice is dangerously like holding a fire cracker between the teeth and biting off the fuse just before the spark reaches the powder. Some psychologists believe cutting passion off abruptly like this, night after night, may induce later sexual difficulties during marriage.

> A boy and girl are interested in each other and so they start to show how interested they are. By this, I mean a date when a boy and girl may be at a drive-in or a regular movie and the boy starts to let his hands roam around and finally the boy and girl get into trouble.
>
> *Girl, 18, New York*

The younger girl who makes a game of heavy necking is a loser. She can't beat this game. How can she keep from getting mixed up in it?

1. She has to want something better from the boy than mere physical contact. She must want to be treated as a person, not a sex. She should expect communication by talk, not by body Braille.

2. Let her check her self-images. Does she think she's so plain or ugly or fat or bent that nobody will ask her out unless she "gives"? Then she's merchandising her sex. She's trading her body for dates.

3. Does she use sex to get even with her mother or her old boyfriend? Sex isn't a weapon.

4. Does she use sex to forget her troubles or escape depression or overcome temporary loneliness? Sex isn't a security blanket.

5. If she sincerely wants to avoid heavy necking, she doesn't dress to tease. She doesn't walk to tease. She doesn't kiss to tease. She doesn't tease. One day some boy will smack her down for teasing. Few people will blame him.

6. She knows that some boys feel they have to *try* to "get" her even if they don't expect to succeed. It's the way some of them are. They're not seriously disappointed when they fail. They're testing her. They're bluffing. Her part of the game is to call their bluff. They'll respect her for it. Some will like her better. They'll come back as friends.

7. She keeps out of hot corners. It's hard to keep a cool conversation going in one corner of a car when every other corner is squirming with action. She avoids crowds which stage this kind of a party. She goes to drive-ins only with boys

who like movies. She says she's having fun and begs to stay when he wants to take her away from a party. She won't go to his house when his parents are gone and won't ask him over to her own house while she's there alone.

8. Considering the consequences, is it worth it?

Thanks to hazily romantic books and movies and sometimes to reports from girls who hope to justify their behavior or to entice others into imitating it (misery loves company?) many of you believe your first experience will be accomplished by the ringing of bells and the trilling of bluebirds. Actually, sexual compatibility frequently takes quite a while to achieve. For many women, it takes many months to reach full sexual satisfaction, this is the most secure of marital relationships. A girl with the added problem of fearing pregnancy, wondering if anybody will find out, worrying that her fiance will change his mind about marrying her—and feeling guilty about the whole thing—is not likely to get or give full enjoyment from sex.

From "Love and Sex" by
Daniel Sugerman and R. Hochstein
Seventeen, July, 1965

Four Consequences of the Sex-Centric Date

Girls want affection just as boys do. They are as interested in sexual stimulation as boys are.

When I decide if I'd like to go out with a guy the first thing I consider is if I'd like to neck with the guy. Also they must be interested in things I am. They don't have to be good-looking but must have a good personality and be able to handle themselves in any situation.
Girl, 17, Milwaukee

All this guy wanted to do was neck and I like to talk sometime.
Girl, 17, Denver

But girls are usually aware of consequences which they, not boys, must suffer.

1. One consequence of heavy necking is either a sudden rush marriage or a child born out of marriage. One out of every three forced marriages ends in divorce. The U.S.

Children's Bureau reports that each year more than a million girls aged fourteen to nineteen have babies and that 10 percent of these infants are born to unmarried mothers.

> A bad date is when any and everything could happen and sometimes it might.
>
> *Girl, 18, St. Louis*

> Most guys—in fact about 80 percent of the ones I know are not virgins. I'd say 50 percent-60 percent of the girls are virgins. People are more lenient with boys because they can't prove otherwise but girls have proof in pregnancy.
>
> *Girl, 18, New York*

Every two weeks or so, a thousand Connecticut girls turn thirteen. During their next seven years as teenagers, very few will die, and relatively few will suffer any serious illness or accident. They face only one major hazard.

If present rates continue, one thirteen-year-old girl in every six in Connecticut will become pregnant out of wedlock before her twentieth birthday.

Nor are Connecticut girls worse off in this respect than girls in other states. The national rate is estimated to be about the same, or perhaps a little higher.

Ruth and Edward Brecher
The New York Times Magazine
May 29, 1966

---

2. A second consequence risked by the girl is disillusionment and deep hurt. She may harden herself against hurt, but this hardening of her personality is itself unattractive and something to avoid. Sexual feelings can be separated from feelings of love. A boy can be sexually attracted to any number of girls at the same time without loving any of them, without ever knowing their names or anything about them. He can genuinely love one girl and still be sexually attracted to others. Girls can also be sexually attracted to several boys all at the same time. But more girls than boys would rather invest their sexual desires together with their feelings, imaginings and thoughts of love all in the same one person.

Suppose now, a girl pins herself to a boy who begs for more and more and more sexual contact, and more and more. He is persistent and persuasive. She finally yields to him. Then he abruptly loses interest. The girl doesn't understand what has happened. She stays home night after night afraid to be out when the telephone rings. It never rings. If she calls him, he won't talk with her long enough to let her learn why his interest changed.

The girl's mistake was to think the boy's feelings were just like her's. They weren't. When the boy's curiosity was satisfied or his sexual tensions were spent, he abruptly lost interest.

> Well, I'd meet boys, and like, you know, I'd . . . I'd *love* them, not just like them. So I stayed with them. And then the summertime, I stayed down here, I smoked a lot, and . . . I didn't just go to *any*body; I really liked these boys. But they used me. Like, I'd tell them I liked them and wanted to be with them, and they'd use me and then say, 'Good-by now, that's it.' And I was very hurt . . .
>
> How do you feel about all that now?
>
> I don't know. I mean, I'm better now, I learned. I don't think everything I did was right, but I don't feel guilty. I mean it's just a misfortune that it went wrong. I learned about things and how to handle them.
>
> But even now, I meet boys and the first thing they say is, 'Come on, let's go to bed.' And I want to show the boy that I like him, but now . . . I don't know what to do.
>
> An 18-year-old Greenwich Village "teeny-bopper" speaking to J. Kirk Sale and Ben Apfelbaum in *The New York Times*, May 28, 1967

A boy who is irresponsible toward lovemaking doesn't usually develop a sudden sense of responsibility when his girl tells him she's pregnant. He runs out or accuses someone else. She takes the rest of her trip alone.

In 1966, the Salvation Army produced a television show describing unwed-motherhood. Six girls from a Salvation Army shelter volunteered to be on the show. Their identity was to be kept a strict secret. During the interviews, they wore masks or talked with their backs to the cameras. Two sets of parents similarly protected also agreed to be interviewed.

But it was impossible to find a single unwed father who would cooperate though dozens of boys were asked.

One boy at last appeared in the producer's office. He wanted to help if he could, he said. But he was worried.

"What if somebody finds out who I am?" he asked.

"No problem," the producer assured him. "You'll wear a mask. We'll disguise your voice by altering the tape."

"But what if my name leaks out."

"Nobody knows your name. I don't know it now and don't want to know it. If I heard your name and accidentally used it in one of my shows, any of my shows, not just this one, you could sue me and my company for millions."

The boy thought a little longer. "No," he decided finally. "I don't want to get involved. I want to play professional baseball. What if I made it and ten years from now this got out? What would happen to my career?"

He left. The show was produced and shown with only the girls and the parents. No unwed father appeared.

3. A third consequence is worry and guilt feelings.

> Sometimes I feel that I'd go with somebody. But I know that when it really comes down to things, I'd be scared witless. I feel that if I went with somebody, I couldn't look them in the face afterward, I really couldn't . . .
>
> *Girl 17, New York*

Girls often judge a date by their feelings the day after. If they wake up and feel guilty or ashamed, they decide they've had a bad night.

Boys are different. Out of four hundred boys describing bad dates, not one boy mentioned day-after shame. All the statements below are from seventeen and eighteen-year-old St. Louis girls:

> A bad date is when I wake up in the morning and feel bad about it. I feel either guilt or anger.
>
> . . . a date that you'd be ashamed to tell anyone about.
>
> . . . one in which you are discouraged from wanting to see the boy again. It makes you feel guilty about something so you feel you won't ever go out again.
>
> . . . one which you would be ashamed to tell your parents about if they asked you what you did on the date. And it makes it worse if they don't even ask you and you know they trusted you.
>
> A regular date is bad when morals conflict and interests differ to a point where bitterness may arise. You end being ashamed of what you did.
>
> You keep worrying about your morals and if you are safe.

4. A fourth consequence is loss of good name. Regardless of a lot of flip talk to the contrary, most boys still want to marry virgins.

Aggressive Necking Is Anti-Dialog

Aggressive necking does more than dodge dialog—it annihilates it. It is a kind of half-hearted rape. There is no

I-Thou here, but I-It. *I* is an aggressor and *It* is a victim. The aggressor tries to force both his conscience and his body on the victim.

The statements below are by girls in New York and St. Louis:

> I hate a date where I have to be fighting a guy off all night. I have to be always on my guard, on the defensive.
>
> I don't like going out with someone I don't like and having it end up in a wrestling match.
>
> You have no fun when a boy tries something on the first date. He gives the girl a hard time as far as trying to make passes. The fella is all hands.
>
> I don't like to be fighting a guy off all night or just at particular times when he's doing this.
>
> I don't like a boy who gets too fresh and likes to travel on a girl.
>
> I hate it when a guy forces himself on me. If a guy wants sex and nothing else and that's all he asks me out for I hate that too.
>
> A bad date would be a date in which all rules and morals are broken by those who have no right to do it. . . . When personalities conflict there are moral conflicts.

Instead of force, the aggressor may use persuasion to achieve his purpose. If a girl quickly and easily yields to a boy, she weakens her chance for mature dialog. She may communicate affection and responsiveness by being an "easy neck" but she also communicates other attitudes and qualities. It may be weakness. It may be fear of not being dated again. She may have a poor image of her own likeability and communicates her lack of confidence by her hunger for any kind of attention.

Trust is very important to a boy's needs. The girl who "sleeps around" communicates that she is attracted and convinced easily. She isn't steady. She doesn't communicate trust, the very basis of a serious and intimate friendship.

The least authentic of all human relationships is that in which boy and girl use each other, where both hide behind lying masks and both treat each other as IT.

Usually the boy is the aggressor and the girl the victim, but to blame boys solely and always is unfair.

The girl may be equally guilty of aggression. Her aggression will consist in provoking, tantalizing and encouraging the boy to make the first move.

In Rostand's play, Don Juan goes to hell and meets all the women he's ever seduced in his lifetime. One by one he discovers he had not seduced them but they him. His hell consists of humiliation and deflation of his ego.

What Do High School Students Think of Virginity?

ARE THEY PROUD OF IT OR IS IT A DRAG?

Students in Denver, St. Louis, Milwaukee and New York City were given the following problem:

> A writer in *Playboy* says our present moral code should be changed. He says: "Our code requires both boys and girls to be chaste before marriage, yet no self-respecting high school or college boy will admit he is a virgin and no smart girl will advertise the fact if she isn't. Therefore both are forced to lie." Assume the writer is being honest and sincere. Answer him.

ON BEING VIRGINS—GIRLS

Look, when I get married I will be a virgin I don't care who knows it, my husband will be receiving a *whole* wife not a half-used, pawed over tramp from the bedrooms of South St. Louis. And if people think less of me because I am a virgin, I say they can go to ----.

Girl, 18, St. Louis

If a boy or a girl isn't a virgin they shouldn't advertise it but they should feel sorry for it, because we all make mistakes. If a boy or a girl is a virgin they should be proud of the fact and not deny it.

Girl, 17, Milwaukee

I agree with him on the first part. The moral code should be changed. It should be strengthened. There is entirely too much sex in today's world. You will rarely watch a program or movie or read a book that doesn't have something to do with sex. There is a time and place for everything. The movie screen or the pages of a pocket book is no place for sex. If sex was put back where it belongs there wouldn't be any need to change the moral code (strength). Any decent respectable boy will not be afraid to admit that he is a virgin. The girl that goes around admitting that she isn't a virgin isn't worthwhile listening to.

Girl, 18, St. Louis

Today if anyone says they are a virgin most everyone thinks it's something to be ashamed of. It's not.

Girl, 17, New York

I agree with the author, pride plays a big role in this area. Many men at time of marriage are not virgins, and the same goes for the girl. Virginity to me does not play a major role in a healthy sex attitude. You don't marry a girl because she is a virgin, you marry her for what she is—not what she was or has done.

Girl, 18, Milwaukee

I think a girl or boy should be proud of being a virgin. Some boys can't help fornication and sometimes a girl can't help herself either. This is a mistake and is wrong but condemning them is also wrong. It can happen to the best of people. If lying is the only way out, it is only to protect their reputation.

Girl, 18, St. Louis

Statistics show that even if a boy isn't a virgin before he marries, he expects his bride to be. The girl(s) he played around with before marriage are left in the cold. So the girl who allows herself to be conned into pre-marriage sexual relationships is asking for it.

Girl, 18, St. Louis

ON BEING VIRGINS—BOYS

The answer to the above statement is this. In my past experience I have known boys who seemed not to be virgins and yet they were. They would always try to be like the others and not be so virginous. They didn't come right out and say that they were virgins, they showed it through the way they acted. Just as no girl wants to tell anyone that they're virgins and yet when they go with boys or are at a party and the boy makes advances towards them they seem to flake off as though they were afraid. This doesn't always mean that the girl or girls are virgins. Sometimes they can do this because they are afraid that if they don't restrict themselves that they might give in to his sex advances or pleasures. And yet no girl goes around admitting that she is not a virgin just as no boy will do it. From the way people are building up the aspect that sex is dirty, I don't think that we'll have to wait much longer before everyone starts to act the same way towards sex. By telling people or making the matter stand out that sex is dirty most kids will try something just because people are saying that it is wrong. If they wouldn't make such a big issue about sex I think our moral code would be much better and we wouldn't have to change it.

Boy, 17, New York

I know a lot of friends who will admit being a virgin and are not ashamed of it, including me.

Boy, 18, New York

I have often said I was a virgin in front of friends and I actually think that they respect you more for it. When it comes to girls being virgins I wouldn't want to marry unless I knew for sure she was a virgin.

Boy, 17, Denver

A self-respecting boy or girl in college or high school should be proud of this fact and if asked about it should give a truthful answer even if it does mean being laughed at by the crowd. I would have more respect for this person than for the person who lied.

Boy, 17, Milwaukee

There are two groups of high school, college girls. There is the group that this writer is talking about. They are teasers, they don't let you know where to take them—to a drive-in or the show. And if you take them to the drive-in they usually get all riled and you know they are just making out to be something they aren't. Then there is the group that is strictly virgin. They don't want to go out with a boy three or four times in a row, and they won't let you kiss them unless they like you. All boys tend to brag, therefore you won't find too many that are happy about their status as a virgin.

Boy, 18, Denver

Communication and Sex Instruction

Parents who themselves are immature can't handle the normal dynamics of adolescence. Such parents don't understand these dynamics. They don't understand the adolescent's biological changes, nor his psychological and social changes either. They are too preoccupied with their own lives to try to understand them.

Such adolescent middle-agers don't communicate with adolescent teenagers on many subjects but communication failure is most obvious and painful regarding sex. One reason may be that they are disturbed or disappointed by their own sexual relationship at this time, so they *repress* talk about sex. But, usually, parents who aren't able to communicate with high schoolers about sex weren't able to communicate with grade schoolers or preschoolers about sex either.

Parents who do not talk frankly and clearly about sex not only fail to give important information, but they give a bad example of communication itself. Some children get the impression sex is something so forbidden it can't even be talked about. Other children feel their parents don't consider them "grown up" (that is, mature enough) to know about sex. They feel they're not trusted and this feeling helps create B.S.I. There is a backlash affect sometimes: children stop trusting their parents.

Thirty senior classes in St. Louis and Milwaukee high schools were asked this question:

What is the strongest criticism or complaint you can offer regarding *your own sex education*?

I didn't get any sex education from my parents. I wish I had because I would probably be closer to them and respect them more.

Girl, 17, St. Louis

I didn't really have any sex education as such until high school. It's something I've never forgiven my parents for and probably never will. My mother never told me anything at all about sex or anything to do with it, not even about the menstrual cycle or anything. I found out everything I know the hard and dirty way.

Girl, 18, Milwaukee

Parents always talk about it negatively. Don't do this . . . But way back in my mind there is always the thought that it is dirty.

Girl, 17, Milwaukee

Too often the Victorian attitude is taken. Or an explanation is started but never completed, or is inadequate. Sometimes things that a person is sensitive about are passed lightly leaving feelings of doubt and confusion.

Girl, 17, St. Louis

Sex was too much of a hushed subject with my mother and just the opposite with my father. Too much of a conflict with ideas confused me on the right attitude.

Girl, 18, St. Louis

I didn't get any sex education from my parents and now when some situation comes up regarding sex I find it very difficult to go to them and discuss it. I always try to work it out myself.

Boy, 17, St. Louis

None given at home or in grade school, but everyone seemed to presume it had.

Boy, 18, Milwaukee

When I was young my parents answered the simple questions asked by a small boy. However when my questions became more difficult they merely handed me a book, I never could read very good and when I read the book I misinterpreted and skipped a lot of the book.

Boy, 17, Milwaukee

My parents never once gave me any explanation of the facts of life. If the parents don't trust their children with this the strongest of all trusts among a family, their total communication will break down. Whenever the sex education at school and other references to sex were made, they acted shocked or surprised making me lose completely my faith and trust in their advice to me.

Boy, 17, St. Louis

My parents were afraid to tell me, they gave me a book and said "read it" this was when I was a freshman, two years after I had learned about sex from an unofficial teacher.

Boy, 18, Milwaukee

I never got any at home, and it was only hinted at in high school (and 8th grade) until senior year. Most "education" was negative: "Don't take out 'bad' girls, be careful not to neck, pet or masturbate" seemed stupid to me when I didn't know what petting or masturbation were.

Boy, 18, St. Louis

My parents were cold when talking about sex.

Boy, 17, Milwaukee

They make you think that sex is a bad thing. They try to make you afraid of it.

Boy, 17, Milwaukee

I had to learn from my friends because my father never bothered to tell me.

Boy, 18, St. Louis

My father didn't tell me; a friend of mine did. I think the father shouldn't tell the son about sex because I think it would be to embarrassing for the father. I think the priest or minister should do it.

Boy, 17, St. Louis

Quiz

PART A

1. Define communication.
2. What are the four typical characteristics of communication?
3. What is dialog?
4. Give both of Reuel Howe's definitions of dialog.
5. Why do we say communication is man's chief means for social becoming? Explain difference between social and personal becoming.

PART B

1. Explain: "Not listening is a form of rejection".
2. What is *seeming* and why is it a problem?
3. Explain the first principle of dialog.
4. Explain the second principle of dialog.
5. In dialog we must be *open*. Open to what?
6. Explain difference as it affects dialog.
7. Explain change as it affects dialog.
8. Define prejudice.
9. How is prejudice different from bias?

PART C

1. Name three ways prejudice affects communication.
2. Name ten ways dialog and communication are different.
3. Describe and explain five barriers to dialog.
4. Give nine rules for engaging in dialog.

CHAPTER FIVE: MAN AND WOMAN: DIFFERENCES AND DIALOG

Dialog and Difference

ON THE CREATION OF WOMAN

In the beginning, when Twashtri came to the creation of woman, he found that he had exhausted his materials in the making of man, and that no solid elements were left. In his dilemma, after profound meditation, he did as follows. He took the rotundity of the moon, and the curves of creepers, and the clinging of tendrils, and the trembling of grass, and the slenderness of the reed, and the bloom of flowers, and the lightness of leaves, and the tapering of the elephant's trunk, and the glances of deer, and the clustering of rows of bees, and the joyous gaiety of sunbeams, and the weeping of clouds, and the fickleness of the winds, and the timidity of the hare, and the vanity of the peacock, and the softness of the parrot's bosom, and the hardness of adamant, and the sweetness of honey, and the cruelty of the tiger, and the warm glow of fire, and the coldness of snow, and the chattering of jays, and the cooing of the kokila, and the hypocrisy of the crane, and the fidelity of the chakrawaka; and compounding all these together he made woman, and gave her to man.

But after one week, man came to him, and said: "Lord, this creature that you have given to me makes my life miserable. She chatters incessantly and teases me beyond endurance, never leaving me alone; and she requires incessant attention, and takes all my time up, and cries about nothing, and is always idle; and so I have come to give her back again, as I cannot live with her." So Twashtri said: "Very well;" and he took her back. Then after another week, man came again to him and said: "Lord, I find that my life is very lonely since I gave you back that creature. I remember how she used to dance and sing to me, and look at me out of the corner of her eye, and play with me, and cling to me; and her laughter was music, and she was beautiful to look at, and soft to touch; so give her back to me again." So Twashtri said: "Very well;" and gave her back again. Then after only three days, man came back to him again and said: "Lord, I know not how it is; but after all I have come to the conclusion that she is more of a trouble than a pleasure to me; so please take her back again." But Twashtri said: "Out on you! Be off! I will have no more of this. You must manage how you can." Then man said: "But I cannot live with her." And Twashtri replied: "Neither could you live without her." And he turned his back on man, and went on with his work. Then man said: "What is to be done, for I cannot live either with her or without her!"

From an ancient Sanscrit text (as quoted by Ernest Crawley, *The Mystic Rose*. London: Methuen & Co., Ltd., 1927, pp 42-43).

Dialog and Man-Woman Differences

The first principle of dialog is that both listener and speaker must be open. They will be open in three ways:

1. They will accept each other where they think and feel the same way.

2. They will accept each other where they are different or unfamiliar or where they disagree.

3. They will accept reasonable change in themselves and each other.

How do we apply the principle of openness to a boy-girl or man-woman relationship such as dating and marriage?

1. AGREEMENT. Both boy and girl must search out the areas where they agree. *Both* must do the searching, not just one. Neither boy nor girl can assume they are open persons if they seek the company only of those they get along with.

The boy and girl must *truly* agree. No "conning." It is dishonest and anti-dialogic if they *say* they agree merely to avoid trouble or to get their way in something else. A boy may agree with everything a girl says in hopes he will get her to feel like lovemaking.

In serious matters, both must agree *verbally*. It is possible for two persons to *feel* the same way but think differently even about their very feelings.

2. CHANGE. Both boy and girl must recognize they are in a process of physical, psychological and social change. If their parents are in middle age, they too will be going through certain physical and psychological changes. Communication is strongly affected by these changes.

Dialog engages the Actual Self and develops the Potential Self. When the body and personality are undergoing extreme changes, the changes themselves become an object of dialog. Open persons are prepared to accept changes in others and in themselves.

3. DIFFERENCES. Men are biochemically, physically, psychologically different from women. They grow and develop differently in all stages of life from childhood to old age. They are usually taught differently by the family and community. Whether society is wrong or right, whether society is changing or not, society still expects different behavior from men and women and holds out different goals and opportunities for them.

All these various differences, in so far as they are present, affect the individual man's and woman's ambitions, fears, loves and anxieties. They will therefore influence their decision-making and communication. If the boy and girl are *open* to each other, they will respect each other's differences.

If the boy and girl respect each other, one will not *force* his needs or amibitions or tastes on the other. Neither will seek satisfaction and pleasure at the expense of the other or by taking advantage of the other. Neither will exploit the other's fears or ambitions or appetites.

Because of training, cultural influences and biological differences single men and women feel their sexual needs in different ways. Their different needs give them different sexual "clocks" and different deadlines. They solve problems and make decisions in different ways. They communicate differently particularly when they communicate about sex itself. When a woman says she feels "sexy," for example, she does not feel what a man feels when he says he feels "sexy." A man's sexual feelings urge him to act, to "invade," to penetrate the woman. The woman's feelings urge her to want, encourage and accept penetration.

Both male and female persons derive pleasure from sexual play. But neither may ignore the child-producing function of their biologies and allow this pleasure to run irresponsibly wild. The boy who tries to excite a girl's sexual feelings

hoping to make her forget her sexual responsibilities does not respect her. He doesn't care for her Potential Self. He treats her as an *It*.

Similarly, the girl who deliberately rouses a boy's erotic feelings by provocative, teasing clothes or talk or behavior merely because she wants a boy's attentions exploits that boy as an *It*.

Prejudice & Co.

If the first principle of dialog is to have open images, the second is to keep them open. Both *I* and *Thou* must tirelessly, vigilantly, honestly keep their images clear and uncluttered.

How do we keep our images clear? By watching for old judgments left over from childhood or adopted from our families. Many of these are left-over prejudices or unexamined biases. They infest our ideas of food, politics and religion, race and nationality, accent, mannerisms, cars and clothes. Of course, not all our old judgments have to be thrown out, just those which were wrong in the first place or which cannot fit new realities. We have to be open to see them as they are today. For example, we learned as children not to talk with strangers. If we followed this rule all our lives, we'd end up knowing only members of the family and marrying persons of our immediate neighborhood. But as we grow older we learn to judge which strangers to avoid and which to trust. In adult society, it is essential that we meet and talk with strangers. Most business is conducted between strangers. The banker afraid to talk with strangers soon goes bankrupt.

Zenophobia, or fear of strangers, is an anti-dialog feeling directly opposed to openness. Not only does it keep us from meeting others but it keeps others from meeting us. The zenophobes put on "cool" masks which disguise their true feelings or even their lack of true feeling. They don't show pain, don't laugh, don't admit liking anything. Their language pretends to be "cool," though frequently the truth is that their vocabulary is severely limited.

Does an intelligent woman represent her Self honestly if she plays dumb with a man to humor his ego?

A common stereotype among girls is that they can get more dates if they pretend to be dumber than the boys they go out with. Their "helplessness" will encourage the boy to assert his "protective nature." Thus, in the Broadway musical comedy, *Wonderful Town*, Ruth sings that a girl will certainly lose her man if she shows him she can start his stalled car by using a bobby pin, that she can out-swim the lifeguard who has come out to save her from drowning, and that she knows baseball like a veteran sports writer.

You've found your perfect mate
and it's been love from the start.
He whispers: "You're the one to who
I give my heart."
Don't say: "I love you, too, my dear.
Let's never, never part."
Just say: "I'm afraid you've made
a grammatical error.
It isn't 'to *who*' I give my heart.
It's to *whom* I give my heart.
You see, with the preposition *to*,
whom becomes the indirect object which
makes the word *to whom* absolutely imperative,
which I can prove
from this very simple chart."
That's a fine way to lose a man.
A fine, fine, fine, fine way to lose a man.
A dandy way to lose a man.
Just be as well informed as he
you'll never hear "Oh Promise Me."
Just tell him where his grammar errs,
then mark your towels "Hers" and "Hers."

Anna Rosenberg Hoffman, former Assistant Secretary of Defense, challenges our old concept of chivalry. She believes it is a male prejudice that women need and want to be protected by men. Contrast her statement below with the excerpt titled "When To Fight" on pages 188-9 in this chapter.

It is fine for men to offer protection to women. But this protection should be limited to their own families. Men have no right to offer protection to any other women. This protection cannot be an excuse for barring women—who have not sought it and do not need it—from taking their rightful place in a working world. A man cannot assume that a woman needs his protection merely because she is a woman.

"A New Look At Women's Work"

Prejudices, biases, clichés, stereotype images, cheap generalizations, false impressions and misinformation are enemies of openness. They are the parasite images which attach themselves to all our important life images. They cause these images to close at the approach of strangers, the "strangers" being new ideas, new experiences and new decisions. They reject changes because every change introduces more strangers. They reject differences because every difference itself is a stranger. The immediate work of Prejudice & Co. is to clog communication or flag it along wrong roads. The ultimate effect is to choke off growth and *becoming*.

The most general, deepest rooted false image affecting the boy-girl or man-woman relationship is one which views women as inferior to men. This opinion has been asserted in our literature and laws for thousands of years. If this opinion is to be believed, women are somehow defective, incomplete creatures, weak in mental ability and generally untalented. The opinion is held by good people and bad, politicians, philosophers and saints among them. It is often held even by women, for it has become part of their self-image. The fact that the opinion is written and printed and published by men does not seem to cast any doubt on its value.

Woman may be said to be an inferior man.

Aristotle

Woman was made to be a help to man, but only a help in generation.

Thomas Aquinas

The Bible says that woman was the last thing God made. Evidently He made her on Saturday night. She reveals His fatigue.

Alexandre Dumas

Can you recall a woman who ever showed you with pride her library?

Benjamin Decasseres

Nature has given horns to bulls, hoofs to horses, swiftness to hares, the power of swimming to fishes, of flying to birds, understanding to men. She had nothing more for women.

Anacreon, 500 B.C.

Most contaminations of dialog can be traced to some false image either of self or the other. A person with a false self-image acts *from an illusion.* His talk isn't authentic because it doesn't represent who he really is.

If a person talks to either a glorified or a prejudiced image of the other person, he's talking *to an illusion* and not the real person. Frequently a false image of the other person points back to a false self-image. The *I-Thou* relationship is impossible if either the *I* or the *Thou* is in any way corrupted.

In the I-Thou relationship, two persons meet and engage each other genuinely as full, individual persons. There can be no genuine meeting if one person does not openly trust the other or has a narrow or low or false opinion of the other. For centuries, men have nurtured such an opinion of women.

A woman talks to one man, looks at a second, and thinks of a third.

Bhartrihari

It is reputed that quite a number of women have had consciences.

James Branch Cabell

Dissimulation is innate in woman, and almost as much a quality of the stupid as of the clever. It is as natural for them to make use of it on every occasion as it is for animals to employ their means of defence when they are attacked. A woman who is perfectly truthful is perhaps an impossibility.

Arthur Schopenhauer

No one but a woman and an Indian ever scalps the dead.

John J. Ingalls

If a man thinks women are by nature inferior and unreliable, he will suspect or discredit their ability to make decisions. If he believes women are by nature untrustworthy, his communication with them can never reach the I-Thou level of dialog.

There can be no full genuine meeting if one person regards the other as immature or undeveloped or incapable of reason.

No woman is a genius; women are a decorative sex. They never have anything to say, but they say it charmingly. They represent the triumph of matter over mind, just as men represent the triumph of mind over morals. . . .

If you want to know what a woman really means look at her, don't listen to her.

Oscar Wilde

Women are only children of a larger growth; they have an entertaining tattle, and sometimes wit; but for solid reasoning, good sense, I never knew in my life one that had it, or who reasoned or acted consequentially for four-and-twenty hours together . . .

Women are to be talked to as below men, and above children.

Lord Chesterfield

Fifty years ago an amendment to the United States Constitution permitted women to help elect their government by voting. Before that amendment, women were legally regarded as incapable of deciding such serious matters. Even today, some modern countries, Switzerland, for example, don't allow women to vote. Up to the 20th century, English law said a married woman had no separate existence of her own.

. . . the very being or legal existence of the woman is suspended during marriage, or at least is incorporated and consolidated into that of her husband.

Sir William Blackstone

There were ancient precedents for these attitudes, as for example, the Code of Manu, an influential and authoritative Hindu law book written between 400 B.C. and 200 A.D.:

In childhood a woman must be subject to her father; in youth to her husband, when her husband is dead, to her sons. A woman must never be free of subjugation.

Dating Roles

While boy and girl usually know the roles they are expected to play in arranging a date, they're not quite sure of what they're supposed to do once the date itself begins. Some girls like to be asked where they'd like to go; some like to be told. Some boys prefer to let the girl choose what to do; some insist on telling. Some girls resent being told. Some boys resent girls who tell them anything. Some girls think boys should lead the conversation and some boys agree with them, but most still expect the girls to do the talking.

> The kind to stay away from is the girl who is always nagging at you telling you what to do, where to go, how to drive.
>
> *Boy, 17*

> When a boy asks you out and doesn't ask what you'd like to do, there's a good chance that you'll just go where he wants. To me a bad date would be to: golf, fish, or go to a park (worse of all).
>
> *Girl, 17*

> I'm not very forward, so I like the guy to be the aggressor.
>
> *Girl, 18*

> I can't stand it if the person you go out with thinks only of himself and what *he* wants to do.
>
> *Girl, 18*

> The girl acts dumb and I have to do all the entertaining and make her mind up where she wants to go and if she doesn't like it, I'm to blame.
>
> *Boy, 18*

The confusion of dating roles comes from a conflict of images. If a girl is annoyed because her date doesn't decide where they are to go, it's because her image is that the man takes care of such things, calls on her, escorts her, sees that she's happy and interested, takes her home, protects her from rudeness and embarrassments. It's the man's *job*, she thinks. When a boy doesn't act this way, he's simply not her ideal of a man.

> Where do you want to go? What a question! If a guy asks a girl out he should have the night planned because usually the girl will agree and feel that the guy has enough power to say and do it.
>
> *Girl, 17*

> The boy I was with didn't talk to me at all except to kid me about not talking. He talked to the other couple whom he knew a lot better than me about old kids they used to know and go to school with. On the way home I did most of the talking because the silences embarrassed me. I always thought that on a date it was the boy's place to start and carry on the conversation!
>
> *Girl, 18*

If a boy expects to decide every detail of a date from the time he meets the girl to the time he takes her home, if he's annoyed at a girl who has ideas of her own, it's because of his images of boy-girl dating roles.

> I can't stand just arguing or letting the girl try to tell me where to go.
>
> *Boy, 18*

Differences and Change during the Dating Period

Adolescence is marked by two periods, the first called *puberty* and the second called *nubility.*

"Nubile" properly applies to both boys and girls, though it is popularly mistaken to mean female breast development. The word "nubile" is from a Latin word which means "ready to marry" and is related to the word "nuptial." The climax of biochemical and physical changes in the nubile male is reached usually at age fifteen or sixteen. At this time the testes have attained maximum growth and are able to produce sperm cells, or sperm, which can fertilize the eggs of the nubile female.

The leading authors of these changes are two remarkable hormones. One of these hormones is *estrogen*, the other is *testosterone.* Males and females produce both of these hormones, though estrogen is known chiefly in women and testosterone is chiefly in men.

Estrogen is present in boys and girls in almost equal amounts until the age of about ten. The amount jumps abruptly in the female at that time while it rises only slightly in the male. In the female the increase of estrogen causes the cells of the uterus and vagina to multiply.

Given signals by estrogen, the female skeletal framework begins to bend in a different shape. The male pelvis is deeper and narrower than the female's so that his legs are straighter and more parallel. The shallow female pelvis broadens, broadening her hips and casting her legs at an angle slightly different from the male's. Her body is adapting to bear children.

The basic difference between the male and female sexes is that only the male produces sperm cells which can unite with egg cells produced only by the female. The glands (testes and ovaries) that produce these distinctly different cells and the organs that release them are called the "primary sexual characteristics." Certain physical changes, such as growth of pubic hair and breast development are called "secondary sexual characteristics." They signal the start of puberty.

How do these primary and secondary characteristics effect decision-making and communication? They have serious effects on both.

Biochemists say that a man is a man because of 5 millionths of an ounce of *androgen,* the so called male "sex hormone," delicately balanced with 2 millionths of an ounce of *estrogen,* the female "sex hormone." They say a woman is a woman because of 50 millionths of an ounce of estrogen balanced with 20 millionths of an ounce of androgen. According to Dr. Ralph Dorfman, a biochemist and hormone scientist:

> A single carbon atom and five hydrogen atoms owned by man is what ultimately marks the man and woman as biochemically different.

The male produces daily a mere 0.0001 of an ounce of testosterone. In its full, original character, testosterone has a life of only thirty minutes. At any specific moment, no more than 0.00001 of an ounce of testosterone is active in the male body. Yet this invisible speck booms and rages in the adolescent boy, deepening his voice, thrusting hair out of his face and body, building red blood cells and muscle, shaping the size and structure of his genital organs, influencing the way he stands and moves.

At puberty the boy's hands, feet, arms and legs race with each other to see which will become bigger first. Shoes are never big enough and he doesn't know where to put his legs. He stumbles. He knocks things over. The nose thickens and the chin pushes out. Face fuzz changes to hair. His voice keeps breaking, tweedling up and down unexpectedly. His face breaks out in pimples. These are all signs of more interior sexual development. They produce drastic and immediate results in his body-image and therefore in his decisions and communication.

The adolescent boy is extremely sensitive to the world about him. Colors, textures, sounds, odors, strike more keenly now than ever before. His hearing is at its sharpest. He is strongly affected by music. His feelings run through seething changes, from snare drum violence to lemon rind humor to a thundering first-awareness of human beauty. Deep underground moods of protest and vicious criticism, futile feelings of love, loneliness and crazy fun times, some moods which skim the egg shell edges of suicide—all these feelings have their dizzying turn during these years.

Some of his old childhood conflicts with his parents are back bothering him. He has to settle them all over again and fit the new solution to his changing bodily condition. He has problems with teachers. He has problems with boy friends and girl friends. He has problems about his future. His new strong sexual feelings keep jumping up to force problems on him too.

Life is a big leathery question mark now. His old basic sense of good feeling has broken up and he misses it. He flounders about trying to find another. But when he tries to share his insights and impressions, his parents and even his old friends seem not to know what he's talking about. Nobody cares about his problems. What does he do? He runs and throws himself into a daydream.

The daydream acts as a kind of de-compression chamber between adolescence and maturity. In it the adolescent can get used to his new emerging self-image and the problems this image is creating. The real world disturbs him so he withdraws from it and creates a world in his mind he would like to have: a mental tree house, a secret island hideaway. At the very center of this daydream world is his daydream image of himself, the Self he would like to be, the James Bond-King Arthur-Super Man from U.N.C.L.E. He has no doubts about his future here. He has it made. He spends his time tinkering with theories, figuring out his life, plotting his revenges. He test-runs new hopes. He is the Invisible Man enjoying all the things the real world denies him. He can date Patty Duke or Nancy Sinatra or Elizabeth Taylor if he wants to. He can be king of the surfers. He can be Bob Dylan singing before screaming girl fans. He can be a Green Beret hero, or a Doctor Dooley, saving lives, almost dying himself but never hurt.

He can do all these things in the privacy of his mind where no one can ridicule or criticize what he's doing or stop him from doing it. Meanwhile, the real world keeps banging against him. Now and then it cracks through his daydream walls and troubles pour in over his daydream self. He is flooded with doubts, fears, conflicts and hostilities. He feels alone, marooned from children and not yet taken aboard by grown-ups. His constant complaint is: "Nobody understands me! Nobody gives a damn."

The tendency to excessive daydreaming fades away slowly. When the adolescent detaches himself from his daydreams and starts re-entry into the world, he is buffeted by a natural turbulence. He has to make new adjustments with friends, parents and teachers as he begins to see the world as it really is. He has new defenses and new ways of coping with this world now. He's rather proud of himself and wants to try himself out. He would like to pit himself against the real adult world. He feels sure of himself and yet feels unsure of himself. He doesn't want anyone telling him what to do or how to do it, but at the same time he has serious doubts about his own ability. He rebels from his parents, yet expects them to back him.

The daydream is natural and a part of normal behavior so long as it doesn't interfere with other natural and normal developments. Some individuals hide themselves in daydreams to avoid real life responsibilities. They make important decisions *in their minds* meanwhile running from necessary decisions right in front of them. They carry on brave, bold dialog with persons—again *in their minds,* while in real life they're too cowardly to open their mouths.

Many adolescent boys like going together in gangs. What the boy himself doesn't have he thinks his gang has. The gang has a kind of style he likes to feel is his style. In this sense, the gang reflects his self-image. The gang may be noisy and rough. Its humor may be loud, brash and exaggerated. Or it may be cool and silent. A gang may form around cars or sports. It may have a uniform—jackets, for example. It has its special heroes and enemies. It has a language of its own based on what its interests are. Within his gang, the young boy may talk and even boast about sex and the gang itself may go girl-watching. But at first members make no actual contact with girls. At this age, girls are taller and more fully developed. They seem to be wiser, too.

One day a member of the gang will admit he is interested in dating a particular girl. His friends jeer and taunt him, but they're secretly jealous. After the date, they are quite willing to believe anything the dater tells them. He may have found the date an uncomfortable experience but he'll say he had a great time.

> When I first started dating I was shy with girls. When I was out with them I was nervous and uneasy. I didn't know what to say or do.
>
> *Boy, 17, St. Louis*

> The first time I ever went out I wasn't too happy about asking the girl—to be precise, I was *scared.* But my boy friend wanted me to double and kept pushing me so when I did ask her, it ended up me sitting on one side of the car and her on the other.
>
> *Boy, 18, Milwaukee*

> Someone set up a blind date for me. She was quite shy. I think she was afraid of me. I was afraid of her too . . . We were both scared and neither of us knew what to say to each other.
>
> *Boy, 17, Denver*

On his first date, a boy is likely to be very sensitive to criticism and easily antagonized.

> A girl said that I wasn't acting my age and I took her home and haven't seen her since, much less talk to her.
>
> *Boy, 17, Milwaukee*

Girls complain that boys are sometimes "coarse," "rowdy," and "too loud" on dates.

> The date just didn't know how to act. He was vulgar. I was embarrassed by him.
>
> *Girl, 17, New York*

He didn't even have the courtesy to talk to me, just goofed around with all of his friends.

Girl, 17, Denver

The boys are "immature," as some girls are keen to note. They have shaky self-images and aren't able yet to act self-confidently.

I have dated a boy a year and a couple of months younger than I am. He was childish and really immature. I felt like his older sister instead of his date. Since the girl is already more mature than the boy mentally, I think she should always date older boys.

Girl, 18, New York

Once I was out on a date and the boy's mother called my house. He had to return immediately. The boy is seventeen and his mother said he was too young to be going out.

Girl, 17, St. Louis

The boy's sudden growth in size and strength directly influences his decision-making. If he's big and strong, he can act out decisions that require muscle to back them up. Strength puts more decisions within his power. Boys who haven't learned to communicate their meanings verbally feel frustrated and may use physical power to force their ideas on others.

The girl's muscular development stops at about age fourteen. Most girls learn early in childhood that they can't win by force. They cope with trouble by using language.

The following problem was given to 729 high school students in Denver and Newburgh, New York:

SOME BOYS YOU KNOW HAVE A SECRET CLUB. IN ORDER TO JOIN, A NEW MEMBER HAS TO BREAK INTO SOMEBODY'S HOUSE AND STEAL SOMETHING. YOUR FOURTEEN YEAR OLD BROTHER TELLS YOU HE'S GOING TO JOIN. WHAT WILL YOU DO?

Response: High School Students	462 Boys	267 Girls
Percent would tell parents at once	11	15
Percent would not tell at all	67	38
Percent try other means, then tell	22	47

The typical girl's answer showed she would *first talk, then tell:*

First, I'd ask his reasons for joining and I'd try to show him it isn't worth it, not only because he might get caught but from the stupidity of the act itself. I'd let him think about it awhile, then, if he still decides to join, I'd inform my parents.

But the most common reaction of the boys was to take some physical action. If a beating failed, they would either inform or give up:

First, I'd beat the hell out of him then I would talk seriously to him and then I would make sure he didn't get out of my sight for a good while . . . Beat up my brother and then get the cops and bust up the punks . . . I'd break his neck . . . I'll tell him not to. If he says he will anyway, I'll probably smash him in the mouth. If he still says he'll do it, I'll tell my folks . . . I'd warn him not to. If he continued, I'd beat some sense into him. Only as a last resort would I tell my parents because they might not understand and if I told them too early they might punish him unduly. . . . I personally would go to the house to be robbed and scare him off. Thus teaching him a lesson without the use of authority or punishment. . . . I would give him as many reasons against as I could. If that failed, I would talk to a priest or counsellor to ask advice. If the advice failed, I would beat him up . . .

Only one girl said she would resolve this problem with force: "I would tell him not to join and if he still persists, I would belt him." Generally, the girls displayed considerable skill and psychological insight. Occasionally they showed the kind of guile that boys find contemptible:

I'd tell him as diplomatically as I could that it was wrong and I'd point out that if he didn't do it because he was afraid of being caught, he might as well do it. I'd also ask who he thought he was that he could go against all his morals and his own self-respect, and what about hurting his parents? What will be the next thing they require? Beating up an old man or a woman? Is this a man's job? If he still insisted, I'd tell him I was going to his parents . . .

I'd tell my brother to show proof that the other members had broken into a house and ask to see what they stole . . . On the night of initiation when the act must be done, the person will be scared because of what he's expecting to do. This is the time to convince him not to do it . . . I'd ask him if he thought it would be all right for me to start smoking, drinking, and some night next week get myself pregnant. I'd tell him that's what he'll end up doing himself and some "nice girl" . . . First, I would tell him (facetiously) "That's *real* cool!", then tell him how stupid a person would have to be to fall for a club like that . . . I'd get somewhere where I could talk to him alone. I'd just ask questions about the club. After he's brought out all the bad in the club, I'd mention a few members who got caught, then just tell him to think about it . . . I would try not to be astounded by the utter ridiculousness of the club. Subtly, I'd try to point up the goodness, character and judgment of his father so that even a fourteen-year-old would want to be like him. He wants to be a man. I'd bank on his childishness to pull out his manliness . . . I would attempt to appeal to his conscience and with the use of irony, shame him. To scold or criticize would get nowhere. He would only laugh at my disgust. I would try to make him hate himself for even thinking of such an idea and when I knew I had hit home I would leave him and go to my room, pray for him and wait . . . Tell him to grow up and start acting his age. If he has to do a stupid thing like joining a secret club anyway, stealing really tops it off nicely. If he wants to run the risk of getting a doolie sentence for a few years, that's ok with me . . .

The Woman

The average daily amount of estrogen produced by the ovary would equal the weight of a thousandth part of a single grain of sugar. The amount produced in a woman's lifetime comes to a scant teaspoonful. The amazing wonder of estrogen is that this invisible quantity helps produce the grand procession of all human beings ever to live in this world, from the first man and woman to all the crawling, walking, flying, praying, warring, eating, building, planting, talking and loving population which covers the globe today.

The breast and nipples are the first to develop in the girl during normal puberty. Breast changes are accompanied by the appearance of pubic hair, which is followed still later by hair in the armpits.

For girls, puberty may start anywhere between the ages of eleven to fourteen. Boys start puberty later, from twelve to fourteen. There is no ruled line for the start of either puberty or nubility. Pubic hair has been noticed as early as eight years or as late as eighteen. Statistical averages are not intended to make anyone feel freakish or abnormal. A statistical freak, for that matter, may in real life be a wonderfully normal human being.

The most dramatic physical event in the adolescent girl's life is menstruation. People once thought it marked the beginning of adolescence but we now know there is considerable adolescent growth before menstruation. Though it happens that some girls don't menstruate until nearly twenty years of age (a very few through some physical abnormality *never* menstruate), usually the first menstrual flow appears at about the age of thirteen. Menstruation is a sign that certain factors, chiefly the hormones, have brought about important adaptations in the female uterus.

It is popularly believed that menstruation is a sign that a female is ready to conceive a child. This is not necessarily true. Two to four years may go by after the first menstrual period before the female releases eggs and conception becomes possible. She reaches nubility when she can conceive, usually at the age of fifteen or sixteen. But there have been cases of fertility and pregnancy *before* menstruation.

Menstruation for some girls is accompanied by cramps as sharp and painful as an attack of appendicitis. Dr. Edmund W. Overstreet, M.D. says that in the week before menstruation about 25 percent of women feel uncomfortable enough to seek medical relief.

Women who don't suffer these acute pains may still feel tension, headaches or some vague sense of depression. Their behaviour may be affected, particularly their communication, but this will depend on their general habits of coping with change. Some women will feel irritable, nervous, moody, ready to scream or cry. Other women feel these things but refuse to react to them. Many, many women feel nothing at all.

Dr. Felix Heald, a specialist in adolescent medicine at the Children's Hospital in Washington, D.C., says:

> Painful menstruation reaches its peak at about the seventeenth or eighteenth year. In general, the data suggests this occurs in 35 to 40 percent of adolescent girls. It probably constitutes one of the major sources of loss of time in the school year.
>
> *Dialogue on Adolescence,* U.S. Children's Bureau

The Girl's Three Escapes from Home

If a girl is too attached to her mother in early puberty, she may have trouble later in her relationships with men. She may become infatuated with an older woman and at such times may become estranged from boys and girls of her own age.

The girl uses several helps in her struggle for independence from her mother.

(1) The daydream is one of these helps. The girl's daydreams are romantic, not usually erotic or ambitious like the boy's. She may have an imaginary romance with some prominent, inaccessible men—Cary Grant has helped at least two generations of girls through adolescence.

(2) The girl nurtures a very close friendship with another adolescent girl. She and her friend will shop together, spend hours on the telephone, sleep and eat at each other's houses. They share many secrets. The friendship is a second means the girl may use to keep from slipping back to dependency on her mother.

(3) The third help is *boys*. Usually at this time the girl discovers her body has a power of attraction. She starts to give attention to grooming and appearance. She uses cosmetics and fashion both as a means to attract men and something to talk about with her girl friends.

The "Weaker" Sex–True?

Some medical scientists think the so-called "sex hormones" may explain why certain diseases take sides with their victims on a sex basis, some diseases choosing men, others running to women. On statistical charts, diseases often pile up separate burial mounds for men and women with steeply contrasting heights. The Arthritis and Rheumatism Foundation reports that ten males for each female are afflicted with rheumatoid spondylitis, a fatal spinal disease. Gout, a form of arthritis, strikes nine men for every woman. But in the case of rheumatoid arthritis, a disease which burns out the body's connective tissue, 80 percent of the victims are women.

The American Cancer Society has graphed similiar mysterious records. Cancer of the lip, tongue, pharynx and esophagus kills four men to every woman. Cancer in the lungs and throat kills almost seven men for every woman. On the other hand, cancer of the breast causes twenty-four thousand female deaths yearly compared with only two hundred and fifty male cases.

The Public Health Service reports that two out of three tuberculosis patients are men. In 1956, 70 percent of TB deaths were males. Haemophilia is almost exclusively a male disease.

Coronary heart disease graphs show the death line for males younger than forty arches ten to forty times higher than the female death line. In all groups, high blood pressure prefers women to men in a two-to-one ratio. At the same time, studies show that women suffer less from high blood pressure's damaging effects.

For many years, doctors tried to explain these uneven piles of statistics piecemeal, one disease at a time. The working and living habits of men got most of the blame. Men had more throat and lung cancer, doctors argued, because they smoked more. Death rates due to brain injuries were higher for men than women because male occupations were more violent. Men worked under greater pressures and strain, therefore they suffered more heart attacks and ulcers. Their hearing suffered more because their jobs exposed them to grinding, roaring crashes and blasts.

Men might be able to avoid lip cancer if they used some kind of lip coating, according to a professor of dermatology in New York City. The lip usually affected with cancer, he asserted was the lower lip, which juts out enough to catch the direct actinic rays of the sun. Men carelessly allowed their lips to dry, crack and split, whereas women shielded theirs by lipstick.

For a long time, piecemeal explanations like these had to do, for science creeps slowly, painstakingly, from one severely limited fact to the next. Some scientists are now beginning to suspect that the death and disease rate difference may somehow be influenced by the different male-female biologies. Their suspicions get some support from child studies where data shows that more males than females are victims of physical disabilities: *three or four boys for every girl suffer from blindness and limited vision; three to four boys to one girl suffer from deafness and poor hearing.*

Stuttering is thought to have psychological origins. A count of stutterers in forty-three cities with populations of over ten thousand revealed *four boys to every girl* affected by this disorder. A later study added three additional points of information: *boys stuttered more severely than the girls; girls "outgrew" stuttering, boys didn't; more boys than girls became stutterers as they got older.*

Differences are not limited to disease and death figures. In 1966 about six hundred adolescents committed suicide. Four hundred were males. For all homicides on record, twice as many adolescent males as females are guilty.

Juvenile delinquency records show a ratio of about *three to four boys for every girl* delinquent.

School records show there are between *three and ten boys for every girl who doesn't learn properly.*

A survey of clinics which specialize in handling problem readers yielded the fact that 78 percent of all children referred to the clinics were boys.

Two-thirds of the children who fail in grade school and are described as "too immature" are boys.

In a survey of learning and behavior problems among 6,026 Maryland children, 919 children were classified as cases severe enough to get special attention. Of this number, 628 were boys and 291 girls. In the same study, 120 children were listed as emotionally and socially immature. They missed school frequently, tired quickly, were underweight and excessively shy. They weren't able to follow directions, couldn't recognize word or letter sounds, couldn't focus attention or hold it for long on the point of any subject. Out of thirty-six first grade children so described, *33 were boys and 3 were girls.*

We are not comparing disease, disability and death records merely to make boys feel inferior or to make girls feel superior to boys. We offer these records because they contradict an error as ancient as it is popular, the notion that women are weak, that men are the master sex.

Investigators have tracked the male human being's weakness to resist stress and shock right back to the womb: two Chicago investigators found that 78 percent of the stillborn fetuses delivered by the fourth month were male.

More boys are conceived than girls by a ratio of about 150 to 100. But according to geneticists, between 15 and 45 male babies die before birth for every 105 male babies delivered safely. On the other hand, very few girl babies are lost during gestation: of every 100 females conceived, 97 to 99 are born. Many doctors are now saying that male weakness is biologically determined in conception itself.

As early as the seventeenth century, researchers noted two kinds of sperm cells, one with small round bodies and another with large longish bodies.

Anton van Leeuwenhoek, the seventeenth century Dutch scientist observed cells through his newly developed high-powered microscope. He wrote in his notebook: "I guessed that one kind were the males, the other females."

After almost three hundred years, Dr. Landrum B. Shettles of Columbia University in New York recently proved van Leeuwenhoek's guess was right.

Dr. Shettles, a camera fitted to his microscope, photographed sperm racing to penetrate female eggs. Millions of sperm, their tails lashing from side to side, moving at a speed of three feet an hour, passed under his lens. His astounding photos show that the small round-bodied cells are speedier and more abundant than the bulkier long-bodied cells. The odds, therefore, favor the small round-bodies reaching the egg ahead of the large long-bodies. Dr. Shettles concludes the round-bodies must be carrying boy-producing chromosomes.

We now know that the sperm cell which drills itself into the female egg to fertilize it is the smallest cell in the human body, only 1/80,000th the size of the egg cell. The adult male normally ejaculates some 400 million sperm cells during each orgasm. They must try to swim six and one-half to seven and one-half inches to meet the egg cell. The egg cell, only 1/200th of an inch across, is the largest cell produced by either the male or the female body. It is mostly inert, non-living food. A baby girl's ovaries contain about 500 thousand egg cells at birth. During her child-bearing years, releasing an egg a month, no more than five hundred eggs will become fully mature.

The sperm cell is a kind of barge loaded with twenty-three chromosomes. In 1959, one of these passengers, identified as the "Y-chromosome," was discovered to be boy-producing. Cells carrying the Y-chromosome are smaller than cells not carrying it. The cells not carrying it are "X-chromosome" cells.

The male sperm decides the baby's sex since it carries the X or Y chromosome. The female egg is always and only X. When fertilized it is either XX or XY. XX means a girl. XY is a boy. The XX egg appears to be fatter and hardier than the XY egg.

The Y-chromosome marks the true biological difference between male and female human beings. It is present in every cell of the male. It is missing in every cell of the female. Some biologists say what this means is that the male and female differ in every cell of their body.

St. Peter called women "the *weaker vessel.*" Shakespeare's Hamlet said, "*Frailty*, thy name is woman." George Herbert in 1640 wrote: "A woman and a glass are ever in danger." Even today we sometimes call women the "weaker sex." What does medical science say about this – true or false? Men are such growly, hairy things bundled with big muscles. They appear to be tougher, stronger, more durable, longer lasting than women. But what are the facts?

Death and disease rates show two things: (1) More men than women are affected by physical and mental stress. (2) When a particular disease attacks both men and women, it is more generally fatal for men not women. Men die, women recover. What the facts suggest is that men may be weaker than women.

Masculine and Feminine

The exact meanings of "masculine" and "feminine" are hard to capture. Their applications are fickle and change with the change of times. Until the Beatles revised an old custom, any teenage boy who wore long hair was jeered at as "feminine."

There are few forms of behavior we can say are always and necessarily characteristic of men or always and necessarily characteristic of women.

For many years in our marriages, the man was regarded as the "head" and the woman as the "heart" of the family. Man was boss. He made all final and important decisions.

Several conditions supported this division of roles. Tradition, training and physical strength made the man family defender and provider. In an economy of plenty or "affluence" such as we are said to have today, a man may consult his wife before he decides where he'll work and what he'll do. But when good jobs are scarce, consulting his wife may seem an empty, even hypocritical, gesture. The man will take what he can get and tell his wife later, "I got a job."

Since family survival usually depends on a man's job and income decisions, it's easy to see why the man should be called the *head* of the family. The head is where decisions are made and the family's most important decision-maker is the man.

Partially by the father's default, mother becomes the family's "heart." Because of work, father is often too busy to clean his children, too tired to play with them and teach them things. Mother is the "heart" also by training and tradition, but underlying this training and tradition is the woman's biology. Her biology makes her want to bear and care for children. Many men would not think of having children if women didn't use psychological forces on them. Many men won't accept the responsibilities of settling down with a family unless they have to; often, only the painful threat of losing the girl he loves is the only pressure that will lead a man to commit himself to marriage.

One Cana group says the characteristics listed below represent male and female roles in marriage:

MAN	WOMAN
FATHER	MOTHER
CONQUEROR	LOVER
CENTER OF INTEREST: THE WORLD	CENTER OF INTEREST: THE HOME
OBJECTIVE	SUBJECTIVE (FEELINGS COME FIRST)
ESSENTIAL-MINDED	DETAIL-MINDED
LOGICAL	INTUITIVE
STEADY	CHANGEABLE
A LEADER	A FOLLOWER
EGOTISTIC	JEALOUS
IMPERSONAL	PERSONAL
PASSIONATE	ROMANTIC
NEGLECTS	NAGS
MAIN WEAKNESS: DISCOURAGEMENT	MAIN WEAKNESS: LONELINESS

The Cana group which publishes these differences does not mean to say that, excepting "father" and "mother," the characteristics always and necessarily describe every man and every woman. They merely say that where a woman can accept herself as a "follower," and accept her husband as "leader" and "conqueror," she can be happy and joyful in her marriage.

But suppose the woman brings home almost as much money as her husband and somehow can't accept herself as a "follower" nor accept him as a "conqueror." If a woman has had straight A's in philosophy all through college, she might chafe if checked from using her head.

Many women, perhaps most women fear independence and the responsibility of final decision-making. They were never trained to assume the role of leadership and are quite content to let their husbands decide everything. It would be harmful and unfair to insist that these women "emancipate" themselves.

There are men, too, who prefer to assist rather than lead. Some men would rather *cooperate* toward a common goal than be sole determinant of that goal. Not everybody wants to be quarterback.

If both married persons want the head-heart division of roles, they can certainly have a happy marriage. But the head and heart division is metaphorical, not biological. Today's married woman has a lot of "head" decisions to make. Today's married man has to have a lot of "heart." Families are smaller today. Many mothers work. The period of motherhood is shorter. Considering these changes and others, a rigorous and literal interpretation of heart and head roles can defeat a marriage.

We think it is wrong for a man to assume he is "conqueror," "a leader," "essential-minded" and "steady" merely because he is a man. There is no evidence that these qualities are printed into the male chromosomes; there is some evidence that they are not.

It is just as wrong to assume that a woman is "intuitive," "detail-minded," "romantic," "changeable" and "subjective" merely because she is a woman. Is a woman doomed to an unhappy marriage or perhaps a series of unhappy marriages, because she is steadier or more objective or more essential-minded than her husband? Must she pretend to be interested in details rather than eternal principles because this is what is expected of her? Must she let her feelings rule her arguments even where she has a better grasp of the central facts than the man?

Each woman will have to find her own personal identity and the role which expresses it, then find a partner to complement it. Every man must do the same.

In the dialog marriage, beyond fatherhood and motherhood, there are no pre-fixed, pre-fab roles to play. Each person desires the full becoming of the other's potential self. Both have the good of the family, the growth and *becoming* of the family as their final aim. This will require a genuine partnership in decision-making and dialog.

Senator Maurine Neuberger of Oregon once complained that she disliked parties where men were separated from women. "In Washington, D.C.," she said, "at party after party you come to that part of the evening where the host or hostess announces that the men will collect in the smoker or den and the women will collect in the dining room or somewhere else. Presumably, the men will discuss problems of the world while the women will discuss problems of the supermarket." The Senator went on to name fifty women in Washington able to hold their own with men discussing problems of government, art, science and religion. She asked for equal conversational rights for women.

But it's hard to dialog with someone who refers to intuition or other mysterious insights and privileged knowledge. It's hard to dialog with a woman who excuses inconsistency, unwillingness to admit error, illogic, missing the point, superficiality or inaccuracy as inborn feminine traits. It's just as hard to dialog with a man who assumes that because he's a man he's "steady," "logical," "objective" and "essential-minded." In dialog these are qualities which have to be struggled for. They are not the natural consequences of maleness.

The passing of patriarchy is in my mind a Christian development that will benefit men as much as it frees women. Mutual equality, freedom, and responsibility will strengthen family life and develop more mature Christians.

The traditional division of work may still hold, but it should be a family pattern freely chosen by both partners. In addition, other patterns of dividing family and community or professional work should also flourish; individual vocations, family needs and community needs may be balanced differently in different families. There is no reason why a complex and sophisticated culture cannot affirm many different vocation patterns, and divisions of work. The satisfactions and dignity of housekeeping and homemaking can be affirmed along with those of professional work. The culture can acknowledge the primary importance of childrearing and family life as well as the values inherent in intensive dedication to the larger community of world and Church.

Now that mankind possesses new economic and scientific knowledge, marriage can more perfectly imitate the larger Christian community with a variety of vocations and works. Today, many married women can respond to their responsibility for family, Church and world in many new and different ways.

Sidney Cornelia Callahan, *The Illusion of Eve,* Sheed and Ward, New York, 1965

The Sex Image

Where does the image of masculinity and femininity come from? It comes from three sources: (1) from the body itself—from the male and female biology; (2) from direct *instruction*; and (3) from the process of *identification*

1. The man's image of himself is dominated by two biological facts: he possesses the seeds of human life and the means to implant them. These facts determine certain of his needs and goals and through them influence many of his actions. They influence his fears, drives and pleasures, his emotional and imaginational life. Whether his awareness of sex is vivid or pale, sex feelings will influence his decisions and communication. The respect he has for his body and responsibility for his power marks both his maturity and moral character.

Three biological facts dominate a mature woman's image of her own sex. The first fact is that she has a vagina, an organ wonderfully designed to be penetrated. The second is that she has a uterus, an organ wonderfully equipped to carry and nourish a human life. The third fact is that she has eggs which, penetrated by a male's sperm, will co-originate a human life.

A psychological fact attends these biological facts: under proper conditions she derives intense pleasure and satisfaction when her organs are being used.

The respect a woman has for her total biology helps establish her dignity, marks her personality and influences her behavior.

Pregnancy is the fulfillment of the mature female organism and as such is attended, under favorable conditions, with deep satisfaction that attend the fulfillment of profound natural impulses. A great deal is said about the need of children for mothers; little is said of the mother's need for children.

Pregnancy is important for a woman because for approximately two hundred and seventy days she walks, talks, breathes, emotionalizes, and psychologizes in the presence of an inner someone. The inner someone is and will be part of her for the rest of her life as it cannot be part of the male. New mothers are frequently accused of having a honeymoon with their infants; but it is a honeymoon that has been going on physically for a secret two hundred and seventy days before birth, and psychologically, the honeymoon with her unborn child has been going on since the new mother was just a little girl.

Lucius F. Cervantes, S.J.
from *The Great Ideas Today*, 1966

---

2. The second influence affecting man-woman roles is through *direct instruction.* The instruction comes not only from parents but from school and neighborhood friends, from books, movies and television. During childhood a boy constantly hears he should "stand on his two feet like a man." (Don't women have two feet and can't they stand on them too?) A girl hears: "Watch the way you sit. It's not *lady-like* to talk that way." A girl is a "tomboy" if she likes to play with guns.

A little boy is praised when he falls down and doesn't cry. A little girl is praised for keeping her dress clean. Her parents may make a big fuss about keeping her covered, that is, keeping her dress down. A little boy is expected to want to be a fireman when he grows up. If he says he'd like to be a nurse or ballet dancer, many families would make a big fuss.

Typically, children are encouraged to play out and practice the parts their parents expect they will play for keeps later in adult life. Most boys play war, dress in military clothes, "drive" toy trucks, use toy electric shavers on their faces. Most girls wear high heels, daub lipstick and mascara over their faces, want their ears pierced. Families differ in their expectations, of course. A little boy whose father is a dress designer may cut out doll dresses. A family may encourage their little girl to play doctor rather than nurse, pilot rather than stewardess. In Israel, where girls serve in the Army, little girls also play with toy soldiers. In Denmark little girls smoke make-believe cigars because their parents smoke real ones. Boys would put on lipstick and mascara too but such a fuss is made against it that they learn not to. Little girls would use toy electric shavers on their chins but they're taught it's something only boys do.

THE STATE OF MIND

The feeling of clear and undeniable sexual identity —which is established in healthy children by the age of two or three—is defective in the potential homosexual. And it is defective because the parents have sexual conflicts of their own. Such parents severely limit their child's chances to develop an accurate understanding of his sexual role. Daily, in countless ways, mother and father indicate "what a delightful little girl their son really is." In extreme cases, the parents may delight in dressing the child in clothes of the opposite sex, in a girl's athletic prowess and stoicism, in a boy's sensibility, feminine qualities and grace. At the same time they condemn the skills appropriate to the sex of the child.

Such parents do not really understand their child's needs nor do they realistically perceive his growth to sexual maturity. Their own immaturity and sexual pathology stands in the way. Fear and anger at the child's budding sexual identity, his sexual assertiveness and curiosity, hamper their effectiveness as parents. Perplexed and confused themselves, mother and father are inconsistent and unreliable when it comes to providing healthy sex images on which the child can pattern himself . . . These parents, to be free of their own anxiety, teach their children codes that are basically in conflict with the biologic nature of man. The children must adopt the codes, and thereby please the parents, because of their need for love, approval and protection. No neurotic homosexual in my experience ever came from a marriage of sexually mature and loving parents.

This point should be emphasized: Children learn effective sexuality from the living example set by their parents. When the example is a good one, the child automatically handles himself successfully in his group. When it is not, and the child fails to get a firm sexual identity, he is bound to have difficulties later on in relating to his contemporaries.

Dr. Howard Davidman
CIBA, April 1958

3. The third influence is *psychologic*. The child wants to be like father or mother or like others who are emotionally

important to him. The adolescent individual also finds someone to *identify* with or against.

The child becomes aware of sex in late infancy. The awareness leads into the start of male and female role values. Typically: little girls identify with their mothers, little boys with their fathers.

Whether a little boy identifies with his father or with his mother at this time will have serious effects in his adult life. If he does not want to identify with his father or some other man, he is likely to imitate the feminine traits of his mother. If he continues imitating feminine behavior during his childhood and puberty, he will probably suffer from a variety of acute personal conflicts. He may become openly homosexual. He may be able to avoid open homosexuality, but may be bothered by such tendencies lurking in his mind and body. They will push to be expressed; he must fight to deny them or keep them hidden.

Sex roles were once much simpler to teach because the sexes were clearly distinguished by customs and laws. A man's job was clearly different from a woman's. He hunted, fought, and did all the heavy work he was physically better suited to do. He wore "work clothes" which identified his trade. The woman watched the home and brought up the children.

But in our culture heavy work is being done more and more by computer-operated machinery. There isn't the same need for muscle. We don't need hunters any more. The fact that women work side by side with men has changed many masculine and feminine manners.

In former times, there was little doubt in children's minds that men and women led different lives. Women didn't vote. Good women didn't smoke or drink. They were rarely doctors or lawyers. A man took his hat off in the presence of a lady. He gave her his seat in the bus or train. He asked her permission before smoking. He told his son: "Ladies first."

The story below was published in 1913 before World War I. Its lessons sound quaint and antique to our ears today:

WHEN TO FIGHT

If you have a chance to do so, boys, it is well to learn to fence and box. To fence or box well you have to give your muscles considerable training, which will make them strong and supple without straining them in any way.

Such training will besides enable a boy to hold his own, should he ever have to do any fighting. For there are times, you know, when even the most peaceable men or boys are forced to fight. I would advise any boy to keep out of a fight just as long as he can, but if he sees a big boy bully a little one, and cannot make him stop in any other way, he should give that bully a good thrashing.

In fact a man's or boy's strength is given him to defend himself against any attack, to fight for his country, and to protect girls, women, children and all those who are weaker than himself.

The other day, I saw in a newspaper that a young woman was kept at work over hours and started to go home alone at ten o'clock at night. It was in a big city, and while she was waiting at the corner of the street for a car, a man stepped up and spoke to her.

This man must have been either drunk or bad, and he must have said something very horrid, for the young woman started back and looked around in a frightened way for a policeman. There was no officer in sight, and the rough man was just going to seize her arm, when another man, passing by, pounced upon the ruffian and gave him the thrashing he so richly deserved.

The newspaper said that the nice man was young and slender, and not nearly so tall and strong as the one he had tackled. But his muscles were well trained, and his indignation gave him the necessary strength to defend that woman.

He did not annoy her by speaking to her, or try to gain her notice in any way, but he held the ruffian down until she had stepped into her car and was out of harm's way. As there was no policeman there, at the time, to protect this lone woman, the young stranger did quite right to interfere and take the law into his own hands, and everybody admires him for it.

Every boy and man should learn to treat every girl and woman just as he would like other men and boys to treat his mother, his sister, or his wife. He should always be ready to protect them from rough men, and to give them any help in his power whenever they need it.

Yourself and Your House Wonderful
by H. A. Guerber. The Uplift
Publishing Company, Pa.

SNACK!

The Dialog Marriage

Images are the valves and filters of communication affecting both what we say and what we hear. They are "valves" because they control how much information we will get from or give to another person. They are "filters" because they influence the feelings we will have toward that person or toward the information we gain or give.

What is our idea of marriage? Does it mean freedom and being on our own? Is it a career? A profession? Is it an annex to the church or the gymnasium?

Does marriage mean having babies—is that all, just having them?

What does love mean to us? How do we picture married life? How does a husband act? How does a wife act? What will we do with our leisure time? How many blocks away from in-laws will we live?

According to a ten year study of marital happiness by Dr. Eleanore Braun Luckey, it is important for a wife to share her husband's image of himself. It is less important for their happiness that the husband share his wife's image of herself. Dr. Luckey found that happy couples tended to agree on estimates of themselves and of each other; unhappy couples did not agree.

Everyone who marries wants a happy marriage. In the happy marriages observed by Dr. Luckey, husband and wife tended to agree in their estimates of themselves and of each other. Dr. Luckey also noted it was important for the wife to share her husband's image of himself. If Dr. Luckey is right, a boy and girl can expect to have a successful marriage if they share certain common images. The discussion and exploration of these images is critically important before marriage.

The most important images in marriage are those of decision-making and communication. Decisions run through all parts of marriage: money management, work, whether the wife works or not, practice of religion, home-making, drink, free time, the education of children, and in-law management.

Applying Dr. Luckey's conclusions to the theme of this book, if a man sees himself as the ultimate decision-maker in the family, he can expect a happy life if his wife shares his view; an unhappy married life if his wife doesn't share his view.

Who is the head of the house? What says it should be the man? What authority says it should not be the woman?

If the man is not the head, does he share authority with the woman? In all things? If so, who has the final word?

Suppose a woman is able to pilot a plane and her husband is not. They're up in a single-engine Cessna. A mountainous thunderhead looms directly ahead, its innocent looking outside fluff hiding vicious winds which could shred their Cessna like a blender.

The husband doesn't recognize this danger. He says: "Go through the cloud."

The wife says: "No, that's a thunderhead."

He says: "Go through. We don't have a lot of gas."

She says: "No. There's turbulence inside. It would wreck us."

At this, the man screams: "Look. Who's boss in this family? You go through the cloud or I'll take the controls and take it through myself."

It's unreasonable to say as a general principle that the husband, merely because he is a man, should always have the last word. The wife-pilot example is an extreme case because it is a matter of life and death. But is this the only kind of decision which is important? Aren't other decisions important too? Isn't the education of children an important matter since it concerns the future life of the child?

Isn't there a better criterion than mere maleness for settling differences of opinion between husband and wife?

Yes, there is. There is reasonableness. All persons must submit to truth regardless of sex, power, wealth, age or experience. Truth from the mouth of a six-year-old child can stand against the powers of the U.S. Congress, the U.S. Supreme Court and the U.S. President joined.

This is the object of dialog: that sweet truth be sovereign, served by both persons who submit themselves to it full-heartedly, not holding anything back for themselves.

The notion that men must rule because they are "more developed, more mature, more reasonable, etc." has done much harm to the true free development of feminine nature. The belief that the male "knows best" automatically because he is the "master sex" is today being tested. It is challenged by women who have been educated in the same universities as men, who are being elected to public office, who are freed from slavery to the home-laundry-bakery-nursery.

Generation by generation, a new kind of marriage is evolving. Success in this new marriage will, as always before, depend on love and respect. But the love and respect will be new too. It will come from a full, genuine meeting of *I and Thou.*

Fourteen hundred and fifty senior high school students were invited to write their own marriage vows: "What do you expect to give up in order to get the married life you want?" The vows they wrote reveal their images of husband, wife and marriage. What do these images indicate? Do they indicate maturity? Some do and some don't.

The images suggest some theoretic matchmaking. First, we will find a boy's image of a husband. We will add to it a girl's image of a husband. If the two images agree or match, we can suppose the couple we've brought together (on paper) can get along together. If there is severe disagreement, very likely they won't be happy together.

We might, for example, expect the couples below to be happy with each other:

> The father should be the head of the house and the mother second in command.
>
> *Boy, 17, New York*

> I would expect him to be the head of the house, make all the important decisions.
>
> *Girl, 18, Denver*

> Everything I would do would be for him. I would be expected to give up all my rights of independence to a certain point. Everything should revolve around my husband and my family. I owe him respect as a human being. A wife always takes the inferior stand to her husband. She owes him obedience. She should try to make him feel important and needed. I owe him love, above all.
>
> *Girl, 18, St. Louis*

> I think the woman should work like a slave for the man. She should cook, sew, cut the grass and do other useful things.
>
> *Boy, 17, Milwaukee*

In the statements below, which girl is likely to be happy with the boy, Girl A or Girl B?

> If my wife loves, honors and obeys me, I think I'd be happy. Love takes care of sex, honor takes care of respect, obey leaves me with authority. I don't think I could live with a domineering woman.
>
> *Boy, 17, Denver*

> A. I expect *him* to be the man of the house, not me.
>
> *Girl, 18, St. Louis*

> B. I'd have to learn to not be so stubborn—this can hurt a marriage. I'd also have to learn to not be so domineering. The husband should rule the house in my opinion. I'd have to get used to not always having my own way. I'd have to learn to adjust to my husband's wants and needs under different circumstances. I'd expect to get love . . . a feeling of security . . . my husband to be kind and considerate—compromising. He might be head of the house but he should consider me.
>
> *Girl, 18, Milwaukee*

The *dialog marriage* is an *I-Thou* marriage. Husband and wife try to meet each other's meanings fully. They respect each other's differences and accept each other's personal changes. They do not force or impose their wills one on the other but accept the sovereignty of truth.

I would have it agreed by both parties, marriage is forever, that all attempts be made for harmony, that differences of opinion be discussed, that neither party should be dominant, that the man earns the money, that the wife raises the children, that nothing be kept secret on purpose from one party, that each party be fully expressed and totally one.

Boy, 18, St. Louis

Each person should be willing to listen to and consider the other's suggestions and ideas. They should agree on basic areas, such as children and money; if they don't agree, they should make a definite attempt at reaching an agreement (these things should never be left unsaid). They should trust each other; keeping secrets even petty ones, can be disastrous. They must make an effort to converse —not light conversation, but serious discussions.

Girl, 18, Denver

I would want a girl to believe in and trust in. Someone who I would comfort and have her comfort me. A girl with the same interests, yet different opinions. I would do my best to understand her and I would want her to understand me. It has been said that no one person can understand completely anyone else, yet I would want my wife to try, and in doing so bring about a more firm trust and strong love. Though all these things might sound like something from a Wonderland, I would want them to someday be a reality.

Boy, 17, Milwaukee

. . . I want to be a wife who can discuss family problems dealing with finance, health, etc. not a scatter-brained floosie. . . .

Girl, 18, Denver

Marriage is a mutual affair . . . Many decisions should be made by husband and wife together. Mutual trust is important. . . .

Girl, 18, St. Louis

It is my opinion that one who expects to get something from the marriage state, one who regards his partner as a means and not an end, will not have a marriage that contributes to the growth of each individual as a human being, consequently, I expect to get very little in exchange. I ask only that I come to a deep and very meaningful awareness of the essence of my marriage partner's individual humanity. But even this is primarily a job for me to do and not a reward for what I've done. Now on to what I expect to give. I expect to give myself, my life and my actions to my partner. In short, I expect to become other-oriented rather than self-oriented. Concretely, this does not mean pleasing each fancy and desire of my marriage partner. It means making each of my actions manifest an awareness of her unique humanness. It means that my actions will be, as much as possible, kind, charitable, and understanding. I expect to give myself.

Boy, 18, Denver

I expect to give everything just about, my time, my help, my personality, my mind, and my body. I expect to direct my aims and my goals along with those of my husband and to constantly try to see him as he is.
I expect my husband to tell me everything he is hoping for and thinking about. He must consider me not only as a woman but a rational human being as well. I expect him to try to be patient and understanding of my contrary nature.

Girl, 18, St. Louis

The first thing I would demand and promise is *honesty.* If both persons are honest their problems will be more easily solved. Also I would let the girl know that I would try to be understanding so that we could solve our problems together. Frequency of sexual relations would be governed by reason and desire of both partners. As far as economics are concerned, together we would work out a budget and try to follow it. We would do as much as we could for ourselves without aid from relatives.

Boy, 18, New York

There should be honesty by both persons. Be able to talk things over before doing important things. I'd want a good job for my husband so he isn't squawking about money matters all the time. . . .

Girl, 18, Denver

Quiz

1. In dialog the principle of openness is applied in three ways. Name and explain each way.
2. What is the most serious and most general prejudice affecting man-woman relationships? Explain.
3. Give five signs which indicate women may not be the "weaker" sex.
4. We all have a complex image of what is masculine and what is feminine. Identify and explain the three sources of this image.

FOR BETTER OR FOR WORSE

Most wives aren't interested in their husbands work and that's why marriages break up. But I make mine talk about his. He's a ticket collector. At first he was reluctant, saying there was nothing to talk about... but I worked out a little game to make it easier for him. All he had to do was remember the different destinations on the tickets he had collected each day and I would ask him about them when he came home in the evening.... It was fun for a while till I discovered he was just making the names up, and it was convincing enough because he has a good knowledge of the places involved but I had no real way of checking....

When I accused him of lying he beat me up.

CHAPTER SIX: THE ENEMIES OF BECOMING

Why Marriages Go Wrong

Some persons argue that it's wrong to drag marital disasters in front of young people. They say it is better to appeal to ideals by describing happy marriages.

But anyone approaching his own marriage decision wants all the facts he can get, pleasant and unpleasant alike. He will listen to honest, pertinent experience. He doesn't want "managed" advice.

Very little is known about the happy marriage. We study marriage failure so that we can see what to avoid. Every failure is the ruin of some couple's grand plan for life. Where did *they* go wrong? How can we avoid their mistakes?

Each cause which contributes to divorce in some way represents a failure in decision-making and communication. By searching through marriage ruins we may discover the means to protect ourselves from these threats to our growth and development, these enemies of becoming.

An air pilot trying to find his assigned landing field doesn't grab for a glossy, colorful travel agency brochure. He studies charts and maps. He talks with the tower. He's grateful to hear there's a beautiful sky and sunny weather but it's bad weather he insists on hearing about. What is the wind's velocity and direction? How much ceiling does he have? If certain conditions at the field have killed other pilots and caused accidents, he'd like to know about them. He doesn't want to hear about successful pilots who have landed safely.

There is a good reason for comparing the marriage decision with the process of landing a plane. There is something so utterly final about both failures.

In the spring of 1965, 1450 high school seniors in St. Louis, Milwaukee, Denver and New York City were given this problem and asked to write their answer:

> A study of six thousand marital breakdowns among Catholics in the Archdiocese of Chicago showed the following nine "causes" to be at fault: adultery, drink,

in-law troubles, irresponsibility, mental illness, money, religion, sexual incompatibility, and temperamental differences. The causes were here presented in *alphabetical* order. Rearrange them, according to your own guess, in the order of seriousness and frequency. Explain your arrangement.

The students' answers were based on personal experience at home, on the marriage experiences of brothers and sisters, on reading, on classroom information and on their own reasoning.

John L. Thomas, S.J., who conducted the Chicago study, said that drink, adultery, irresponsibility and temperamental conflict accounted for 80 percent of the marriage breakdowns leading to divorce.

A remarkable fact is that the student guesses very nearly matched Thomas's scientific findings.

THOMAS

1. Drink
2. Adultery
3. Irresponsibility
4. Temp. Differences

BOYS	GIRLS
1. Irresponsibility	1. Irresponsibility
2. Adultery	2. Adultery
3. Drink	3. Drink
4. Temp. Differences	4. Temp. Differences

The student opinions are interesting because they suggest which threats they fear most regarding their own marriage and family. The opinions thus reflect an image of marriage which will probably influence their own choice of marriage partner.

It starts off with trouble because of religious matters, that night he finds out she's incompatible. Then she starts yelling about money that she needs and her in-laws tell her what a loser you are. You start to drink and you develop temperamental differences which finally lead to driving the poor guy to the nut house because he developed a mental illness. And later she goes out and commits adultery with some big jerk.

Boy, 18, Milwaukee

Many people today enter marriage thinking it is a "bed or roses." They think all their troubles will end once they are married. They

don't know the first thing about managing a house or a budget or raising a child.
In-laws are another source of trouble. They are constantly trying to improve someone or something in your home and are forever against modern ways of doing things. I think money is the least of the problems. People can always make a go of it even if they have to borrow.

Girl, 18, New York

The Alleged Causes of Marriage Failure

Here is Thomas's distribution of causes according to factors involved in the breakdown of marriage:

CAUSE OF DIVORCE	PERCENTAGE
Drink	29.8
Adultery	24.8
Irresponsibility	12.4
Temperaments	12.1
In-Laws	7.2
Sexual Incompatibility	5.4
Mental Illness	3.0
Religious Differences	2.9
Money	.8
Unclassified	1.7

Here is how the students estimated the threats against successful marriage:

	RANK	CAUSE	
Boys:	1.	Drink	(177)
	2.	Irresponsibility	(171)
	3.	Adultery	(159)
	4.	Money	(118)
	5.	Temp. Differences	(111)
	6.	Sexual Incompatibility	(101)
	7.	In-Laws	(74)
	8.	Religion	(80)
	9.	Mental Illness	(23)

	RANK	CAUSE	
Girls:	1.	Irresponsibility	(218)
	2.	Drink	(194)
	3.	Adultery	(170)
	4.	Temp. Differences	(133)
	5.	Sexual Incompatibility	(116)
	6.	Money	(115)
	7.	In-Laws	(108)
	8.	Religion	(85)
	9.	Mental Illness	(18)

It's not always possible to know why a marriage breaks up. Thomas says:

> By the time marriage partners have reached the point at which they decide to separate, a whole series of incidents in word and action have accumulated. These are often recited to the counselor with no attention to time, casual sequence or relative importance. What is the real root of the trouble? At times, the couple themselves do not know. At times, even the experienced counselor is unable to discover the real source of disharmony and has to sum up his opinion of the case under the empty, cover-all of "incompatibility."
>
> John L. Thomas, S.J.
> *The American Catholic Family*

ENEMY OF BECOMING #1:

Drinking and Broken Marriage

We frequently hear of people dying from too much drinking. That this happens is a matter of record. But the blame almost always is placed on whiskey. Why this should be I never could understand. You can die from drinking too much of anything—coffee, water, milk, soft drinks and all such stuff as that. And so long as the

presence of death lurks with anyone who goes through the simple act of swallowing, I will make mine whiskey.
W. C. Fields.

At a certain point of time after high school, one particular bottle of beer or one shot of whiskey starts some young men and women on a career of problem drinking which will make their married lives reek of disappointment and suffering.

> Drinking is a major sickness in America today and it is very hard to live with a drunkard. I am from a family like this, and I know.
>
> *Boy, 17, Milwaukee*

> He [my father] gets a few under the belt and does things he would never do otherwise.
>
> *Boy, 18, St. Louis*

> My dad drinks, and he just leaves when he does, just walks out after supper and will come home 10 or 11 o'clock. Who he's with no one knows. Goes to Mass on Sundays and just Mass. He buys just some of the food and pays bills when he feels like it. Wants to have sex relations almost every night; mental; he gets the weirdest ideas, about where we've been and what we are doing; never trusts us. Relatives just say or tell us he needs help but they won't help us make him get help.
>
> *Boy, 17, Milwaukee*

Liquor is expensive. Often the money that buys it was budgeted for something else—the rent or payment on a car or bank loan. Consequently, drinking is often associated with non-support. Money needed for the house is thrown away.

> Nobody knows how much he makes except the tavernkeeper who gets it.
>
> *Boy, 17, Milwaukee*

> I don't have a father on weekends. On Friday night he moves over to the tavern and that's the last we see of him till he goes to work Monday morning.
>
> *Boy, 17, St. Louis*

The high school students in our survey connected drinking with destruction and waste.

> Every week he breaks something else . . . last week he tried to save himself from falling and tore the new shower curtain down rod and all . . .
>
> *Girl, 16, New York*

Thomas says that drinking usually attacks a marriage paired with some other form of bad behavior: alcohol and cruelty or violence, alcohol and irresponsibility, alcohol and adultery. According to Thomas' study, drinking among laboring class workers was usually accompanied by beatings and abuse.

> I told my brother he's a boy and he should try to stop him but my brother's afraid of getting a beating too. We're both afraid.
>
> *Girl, 17, Milwaukee*

> The parents are constantly fighting . . . When a person drinks it causes the rest of the family to lose respect.
>
> *Girl, 17, St. Louis*

> Drink seems to release all of our tensions. So it would seem only natural to release the rest of our troubles at the wife and kids.
>
> *Boy, 18, New York*

> If a married couple with children drank and stayed at bars all hours of the night, this is very hard on the children. No wonder they turn out to be juvenile delinquents.
>
> *Girl, 18, New York*

> I can't stand to see him hit my mother. He's making me hate him.
>
> *Boy, 17, St. Louis*

In 1962 the American Humane Association found 662 newspaper reports of parents who beat, burned, drowned, stabbed and suffocated their children with weapons ranging from baseball bats to plastic bags. Most of the victims were under four; 25 percent died. If all such cases were reported, say some experts, the total would reach 10,000 a year. Many doctors suspect that more U.S. children are killed by their parents than by auto accidents, leukemia or muscular dystro-

May I take this Opportunity to

phy. In most cases parents when charged with assaulting their children plead they were under the influence of alcohol.

Drinking is alleged to be the No. 1 cause of Catholic divorces, according to the Chicago study directed by Thomas.

> Failure was admitted only when, through the refusal of one of the spouses to fulfill an essential marital role, the marriage lost all semblance of a real union and deteriorated to a mere external form or became an instrument of physical and moral harm to the partner and the children.
>
> John L. Thomas, S.J.
> *The American Catholic Family*

Excessive drinking accounted for 30 percent of all divorces.

> Drink, to me, seems to be a problem with just about everyone these days. They get liquored up and come home and beat or shoot the family. You can't blame them for getting out of there.
>
> *Boy, 18, St. Louis*

Though today the number of women alcoholics is rising, in Thomas's study almost always men were the offenders. They were not victims of the *disease* of alcoholism. Most were "weekenders"; heavy but periodic drunks.

xtend to you my Most Sincere Wishes
for a Very Merry Christmas,

Drinking will break up a marriage unless the sober party realizes it is a disease.

Boy, 18, Denver

Drinking is a disease and it's hard for someone who doesn't have this problem to be understanding and patient.

Boy, 17, Milwaukee

The Chicago data showed Thomas that wives seemed reluctant to give up their drinking husbands and did so only after putting up with considerable suffering. It was obvious from the number of children and length of time the marriage had lasted that affection was slow to die.

Concerning drink, how many people still live with their marriage partner even if they are drunk for the simple reason of the children's sake?

Girl, 18, Milwaukee

Psychologists now say that women who marry and hang on to drunkards do so because of unconscious needs. But the wife who hung on for the sake of "appearances" or for the children is not so common as formerly. The growing economic independence of women in marriage is shortening their tolerance and will to endure old-fashioned suffering. Some girls clearly indicated that romantic patience is running out.

A person may come home night after night drunk and beat his wife, ruining his and her lives. Drinking is hard to put up with and even harder to fight against.

Girl, 18, Milwaukee

and my Prayers that
Health, wealth, & Happiness attend you

It is very hard to live with someone who is always drunk. Sometimes you may be risking your life.

Girl, 17, St. Louis

No matter how strong the love is, this is one fault which cannot be endured even by love. Mainly because it leads to so many of the other causes.

Girl, 18, Denver

One might become fed up and forget about the "for better or worse" statement of their vows.

Girl, 17, St. Louis

I myself wouldn't put up with an alcoholic husband.

Girl, 17, New York

Alcohol: Stimulant or Depressant?

> A man who overindulges, lives in a dream. He becomes conceited. He thinks the whole world revolves around him—and it usually does . . .
>
> I'd hardly quit liquor; before I got the d.t.'s I'd see little men with whiskers and high hats, sitting on skulls, and they'd charge me. They almost got me one afternoon.
>
> W. C. Fields

in the Coming Year!

Many of us think alcohol is a "stimulant." But medical scientists say the opposite is true; alcohol acts as a *depressant*. The word *depressant* or *depress* as used in medicine doesn't necessarily refer to sadness or gloom. Rather, a depressant muffles and weakens an organ so it doesn't perform at its best, top condition. Alcohol blurs our vision, slurs our speech, makes us mispronounce words or confuse the order of what we want to say. It upsets our balance and timing. Late stage alcoholics feel only the barest wisps of sexual desire and may feel them as rarely as once or twice a year.

DEPRESSANT: In psychiatry, a morbid sadness, a dejection, or melancholy. It is different from grief, which is a realistic sadness and is proportionate to what has been lost. A *depression* may be a symptom of any psychiatric disorder.

Alcohol *seems* to be a stimulant because of its depressant effect on that part of the brain called the *inhibitory center*. In general, the inhibitory center is the area of the brain where our commands, judgments, prejudices, distinctions and the like are preserved. They are called "inhibitions" when they block our acts.

Not enough is known about the brain to mark off strictly what or where the inhibitory center is. In general, it is said to be in the outside layer of the brain, the gray matter known as the *cortex* or *cortical* area, which is thought to be the center of human consciousness.

Picture a jumble-tied knot where all the wires of an electronic data-reporting instrument meet—that's the cortex. Here, in this puzzling network of coils and loops, messages are impressed, somehow, as if on high-speed film or recording tape; all the lessons, threats and promised rewards we've learned from childhood to this very moment.

Every judgment or decision we've ever made, every link between noun and verb has flicked through here and been counted. Here are arithmetic tables, football and basketball plays. Here are school rules, traffic regulations, Mass rubrics, the "sense" of time that buzzes inaudibly when we should stop what we're doing and head for home. Here is the law of Moses and the life of Jesus and the Scout Code. Here all the warnings that ever impressed us: *Don't stare. Don't touch. Don't talk to strangers.*

In the cortex, too, are phantom sexual symbols—chips of language, images, bars of sound and music that can stop or start our erotic emotions. Here are the lines that control kissing and petting. Here, also, the religious principles, codes, and bargains that rule and regulate our lives: *I pledge allegiance to the flag of the United States of America . . . in sickness as in health till death do us part . . . Thou shalt not commit adultery.*

R. O. BLECHMAN

Promises, agreements, verbal contracts, resolutions and pledges can be lost if they have shallow roots. The first to go are often those which we resented or which were hard to absorb or grasp by the mind. If the obligations and implications of any bargain or agreement are expected to last, they have to be clearly understood. We have to grab them good and hard. We have to make them a habit. If a person has a weak grip on his marriage vows, he'll find he can let go when he's drunk. If he resents the responsibilities of married life, alcohol makes it easy to ignore them.

Alcohol vs. Becoming

ALCOHOL DISSOLVES THE TWO MOST INTIMATE AND CRITICAL MEANS OF GROWTH IN MARRIAGE: DECISION-MAKING AND COMMUNICATION.

ALCOHOL AND DECISION-MAKING

Drinking affects each of the six rules for decision-making:

THE RULES

1. Don't run from a decision.	When troubles mount up, drink spares them of reality. Drinking is due to immaturity and the inability to accept trials. It's an escape to nowhere.
2. Avoid important decisions when fatigued or when over-exhilarated or depressed.	We usually act on information as reported by the senses. Our decisions are bound to be faulty because the senses are not reporting accurately.
3. Write down the main facts.	Drinking endangers ability to distinguish important facts from minor, accidental facts.
4. Separate minor and major goals.	Drinking confuses and obscures goals.
5. Collect and grade advice.	Drinking interferes with listening. The first good advice the drinker hears is "lay off liquor." He ignores it.

6. Decide what's in your power to do. — Alcohol's depressant action directly affects a person's power to perform. At the same time it encourages the drinker either to over-estimate or under-estimate his abilities.

Alcohol and Communication

There's no doubt that a little alcohol may help conversation. Dr. Alfred Kinsey confessed small amounts of alcohol helped some people reveal their sexual and personal affairs to him freely. We aren't talking about "a little" alcohol nor about normal social drinking. We're talking about a person who uses alcohol to evade the troubles and responsibilities of dialog.

An excessive amount of alcohol affects communication in these important ways:

1. It violates the first principle of dialog. Its effect is not to open but to *close*.
2. It distorts the drinker's *image* of himself and of the other person. In dialog we attempt to correct false images and reject masks which hide or project different imaginary identities.
3. It distorts the *information* we need to dialog properly.
4. It distorts our personal *expression*.
5. It affects behavior which in turn affects dialog.
6. It isolates *I* and prevents *I-Thou*.
7. Its end effect is to reduce *I* to *It*. No one can relate as a human person in a state of unconsciousness.

> Drink—shows selfishness—desire to be alone away from family—desire for escape.
>
> *Girl, 18, New York*

> One person tires of the other and the bottle is an out.
>
> *Boy, 17, St. Louis*

No one has yet been able to lift the roof off an inhibitory center in the brain and watch just what alcohol does there that makes us let go of our inhibitions. But we do know alcohol hampers or depresses the function of the nerve cells in the cortex. In small amounts it may quiet anxieties, reduce sensations of fatigue, undo knotty tensions. But small amounts may lead to larger amounts.

The more alcohol we take in at a given time, the more the brain is penetrated. Gradually we hear and smell less and less; our skin fails to report small changes of heat or cold, wet or dry, smooth or rough. Our vision gets blurred and swirly.

The brain that has alcohol sitting on its cortex may still get information sent in to it from the senses, but the perceptions wobble in distortedly—the more alcohol, the more wobble. Reactions to this wobbly information are usually slow and distorted too. Distinctions are fuzzy, judgments run together.

One immediate consequence is that the individual's usual images become distorted. A drinking man may think he sees Miss Teenage America in some hag who should be riding a Halloween broom. A drinking woman may find a dull man's conversation exhilarating. John Phillip Sousa's marches may sound like Debussy's Prelude to "The Afternoon of a Faun."

The steady creeping freeze of his senses insulates the drinker from the outside world as effectively as a numbing anesthetic. He sniffles and cringes or he boasts or gets belligerent. Large amounts of alcohol usually have a jamming effect on self-criticism. It allows a shy, self-controlled person to wear lamp shades at a party or start an argument with his boss. On the other hand, it may release a flood of self-criticism and turn a confident person into a self-accusing, soggy, apologetic blob of self-pity.

Finally, like a bad television picture, the images begin to roll and flutter. There's a first brief shock of intense light and the picture disappears entirely. He's out cold.

> He [father] just blacks out. Nobody but nobody can talk with him he's so gone. If you need money for lunch he can't hear, just sits and looks at TV without seeing.
> I can never have friends to my house because he's always there. He gets loud and embarrassing. He argues with my date instead of talking with him.
>
> *Girl, 18, St. Louis*

Student Drinking

Some students said that teen-age drinking was a widespread and serious problem:

I think drunkenness and alcoholism is on the rise among teen-agers who will be the married couples of the future. This is why I make it No. 1.

Boy, 18, Milwaukee

I think drinking is also going to be an even greater cause in the next ten or fifteen years because of the great number of teen-agers today who drink.

Girl, 17, New York

A report recently published by the Rutgers Center of Alcohol Studies contradicted these gloomy views. The report showed that out of nearly two thousand junior and senior year public high school students in Michigan only about 1 percent reported "frequent" drinking, 23 percent said they drank "occasionally." Many "occasional" and some "frequent" drinkers drank in the company of their parents at home.

High school students don't regard drinking as a dangerous or bad practice. The following question-problem was put to 870 Denver and New York high school students:

> YOUR PARENTS LET YOU AND YOUR BROTHER GO TO AN OUTDOOR PARTY AT NIGHT BUT PUT YOU ON YOUR HONOR NOT TO DRINK ANYTHING ALCOHOLIC. YOU ARE SUPPOSED TO WATCH OUT FOR EACH OTHER—BUDDY SYSTEM. THEN SOME OF THE BOYS START CHUGGING BEER AND YOUR BROTHER JOINS THEM.

Response: 267 Girls—*Would You Report Drinking?* 68 percent "NO".

SAMPLE ANSWERS:

I don't think it is such a crime to drink at a party. . . . No. I would feel he was probably so "high" he didn't realize what he was doing. . . .

Response: 662 Boys—*Would You Report Drinking?* 87 percent "NO".

SAMPLE ANSWERS:

No. I wouldn't rat because I would start chugging too . . . Chugging beer is fun . . . If your brother or you get busted for it, that's the chance you have to take. . . . I feel there's nothing wrong with drinking as long as no violence results . . . A person shouldn't be forced not to drink if he's willing to die for his country. . . . Many parents freely and willingly let their children drink. Mine are included.

A poll conducted by the Institute of Student Opinion in 1964 asked 7000 students in all parts of the country this question: "How would you rate drinking as to the amount of danger it creates or would create to your good health, good morals and future happiness?"

HOW TEENAGERS LOOK AT DRINKING

	Would Not Harm Me in Any Way	Might Be a Slight Danger, but Not Much	Would Be of Serious Danger to Me	No Response
Drinking Alcoholic Beverages at Home				
Total:	8.8	23.8	65.2	2.2
Boys	10.5	24.9	61.5	3.1
Girls	7.2	22.8	68.7	1.3
Boys: 12 to 14	4.8	15.2	76.3	3.7
15 to 16	9.0	25.8	62.7	2.5
17 and over	16.6	31.7	50.2	1.5
Girls: 12 to 14	2.6	13.6	82.0	1.8
15 to 16	6.7	25.3	67.1	.9
17 and over	12.8	29.5	56.3	1.4
Drinking Alcoholic Beverages On Dates or Out With Friends				
Total:	3.2	13.5	81.1	2.2
Boys	4.7	17.6	74.7	3.0
Girls	1.7	9.7	87.3	1.3
Boys: 12 to 14	3.3	10.6	83.2	2.9
15 to 16	4.0	15.7	77.8	2.5
17 and over	6.6	25.0	66.5	1.9
Girls: 12 to 14	0.7	5.6	92.2	1.5
15 to 16	1.8	10.0	87.2	1.0
17 and over	2.7	13.8	82.1	1.4

Scholastic Magazine

Student Attitudes Toward Marrying a Drinker

The following statement was made by Sara J., 26., to a marriage counsellor in Detroit, Michigan:

"I knew Sam drank too much and that he ran around with women. But I thought if I prayed hard enough and loved him enough he would change after we got married. I decided to marry him and trust in God."

1,100 high school seniors—650 girls, 450 boys, were asked to comment.

GIRLS: *6 PERCENT OF THE GIRLS THOUGHT SARA J. HAD MADE A GOOD DECISION IN MARRYING SAM.*

I think this lady loved Sam and she wanted to help him. This is the right decision. Everyone has some faults. Sam happened to drink and run around with women. But this lady put her trust in God and I hope it worked out.

Marriage is a big step and though I think its un-christian to be practical sometime, if you don't want to get hurt, you have to be practical, and not presume things. There are a few people who don't care about getting hurt, and would marry Sam anyhow just because they're needed. Of course, the children might be a problem but a wise mother could almost make up for a lax father.

BOYS: *10 PERCENT OF THE BOYS THOUGHT SARA J. HAD MADE A GOOD DECISION.*

This is the only good reason I've heard yet. If you honestly pray and love and believe God will help, I think I can safely say, "He will help you."

If this woman has the faith and courage to put up with Sam and trust in God and herself to change Sam, I think she did right.

Through the help of God and the A.A. a successful future may still be in store.

I think this woman has more faith in God's effect on people than most of us. I would never marry a woman knowing this, and to me it sounds like this gal is lost and needs someone to confide in; but praying, to my mind, will not help this man in most cases stop drinking and flirting. Maybe, however, she could be just what he needs—someone to tie him down and, by her example show him what love really means. I say this, because I know of one couple, in particular, for whom this problem was solved after five or six years of marriage.

45 PERCENT OF BOTH GIRLS AND BOYS AGREED IT WAS WRONG TO EXPECT MARRIAGE TO REFORM A DRINKER.

Never marry a man to change him. Marry him because he is what he is.

When you choose a partner, you don't choose him for his past or future, you marry the man you know.

Marriage is to live a life of love and sharing, and children, not to reform. Marriage is no place to mature.

If a guy is a slob before he was married chances are marriage isn't going to change him.

Marriage solves no problems, it only creates them.

In a relationship the persons are only transformed into something else because of previous abilities and capabilities. No one ever is totally transformed.

Marriage is no place to reform a man, he is what he is.

You have to trust in Sam, too! If you can't, it won't work. Love includes trust. But you cannot desire to make a *basic* change in a person—you must respect someone before you can love them.

BOYS: *90 PERCENT THOUGHT SARA'S DECISION WAS FOOLISH.*

It is very touching to have such blind faith, but also very unrealistic. Just because a man takes a few marriage vows one morning doesn't mean he is going to start a new life. . . . Well, I admire this woman's faith but not her intelligence. These kind of guys never change. . . . This marriage was on the rocks before it got out of the water. The only thing I can say is, at least there may have been some love there . . . her prayers to God doesn't change the taste of liquor . . . God can help you but that's a little too much to ask don't you think? . . . It will never work out, you just cannot change a person's ways by hoping and praying. You have to face reality. . . . This sounds like a wish in a fairy tale . . . Faith put in for a hopeless cause is loss of faith itself . . . Pick the best marriage partner—then trust in God. . . . With your faith in God you should have become a nun, then you could have prayed for him all your life. . . .

GIRLS: *94 PERCENT THOUGHT SARA'S DECISION WAS FOOLISH.*

Congratulations St. Monica. What a dreamer. . . . God can help but I still think it would be a lot better if Sara waited to see if Sam wanted help. . . . Are you kidding?—once a rat—always a rat. Just because you marry the guy doesn't mean he'll turn into a saint. All the loving and praying in the world can't change a guy like this. . . . This might be good if a person prays for a pretty long time but what happens in-between or while she is waiting for him to change? . . . Everybody is always leaving things up to God. If you were running and saw a bridge coming would you run right off and wait for God to come swooping down from the sky like "superman" to save your life? . . . Sure trust in God—and make up in hell! You cannot live by sitting on your ______. . . . Well honey, be prepared to be alone . . . Sara was trying to be a savior instead of a woman with good common sense. . . . She isn't being fair to the children he must lead. . . . Personally I believe Sara enjoys being cheated on and treated cruel, for she knew what she was getting into and by blind trust in God was her cowardly way of shunning responsibility.

THE THIRTEEN STEPS TO ALCOHOLISM

1. Increasing consumption of alcohol.
2. Digestive disturbance—early morning nausea—loss of appetite.
3. Drinking that interferes with the conduct of one's normal life.
4. Missing time from work or duties because of drinking.
5. Drinking more than intended or getting drunk when not intended.
6. Necessity for a drink—particularly at a certain time of day—using alcohol as a drug rather than a beverage.
7. Constant thinking about drinking.
8. Drinking becoming more important in life.
9. Refusing to admit what is obviously excessive drinking.
10. Drinking in the morning—or on awakening.
11. Making alibis for drinking.
12. Family dissension and quarrels because of drinking excessively.
13. Getting drunk alone.

American Medical Association

Quiz

1. Is alcohol a stimulant or a depressant or both? Explain your answer.
2. Describe how alcohol affects each of the six rules of decision-making.
3. State six of the American Medical Association's thirteen steps to alcoholism.
4. Describe seven ways alcohol affects communication.
5. Describe two ways alcohol is related to adultery.

ENEMY OF BECOMING #2
Adultery

Whosoever looketh on a woman to lust after her hath committed adultery with her already in his heart.
Matthew v, 28 c.75

It shall be considered adultery to offer presents to a married woman, to romp with her, to touch her dress or ornaments, or to sit with her on a bed.
The Code of Manu, viii, c. 100

No wrong strikes closer to the heart of marriage than does adultery. The Church recognizes adultery as grounds for separation. Students thought adultery was repulsive. Could they forgive? Some could, some couldn't.

It is something that just can't be forgiven.
Girl, 18, New York

No woman would want to live with her husband if he had something to do with another woman.
Girl, 18, Milwaukee

Drink and adultery—these seem to me faults which true mature love should be able to forgive if the other party is truly sorry. If not . . .
Girl, 18, New York

Adultery can't usually be forgiven by one's mate because he or she will always ask "Am I as good as he (she)?"
Boy, 18, St. Louis

Most persons are willing to forgive the first time. But a second time usually ends the marriage.
Boy, 17, Denver

It is that the partner went against you directly, intentionally. It is hard to forgive something like that.
Girl, 18, Milwaukee

The sin of adultery is bad but when it is committed by the man it is usually forgiven (after a certain amount of bribery).
Boy, 18, Milwaukee

Adultery—this would make the feeling between partners quite different and would cause great heartache and hurt and a great loss of love and respect for one's partner.
Girl, 18, St. Louis

Marriage is a unique sharing of privacy. Adultery introduces an outsider, an intruder to the personal and, intimate communication of married life.

It is hard for a husband or wife to trust his spouse after this.
Girl, 17, Milwaukee

I could tolerate many things from my husband but not being untrue to me. It's very hard to love a guy you know had a relationship with another woman while you were married.

Girl, 18, Denver

I don't think there is anything more disheartening than the belief that your love does not satisfy your husband and he must find other means of happiness.

Girl, 18, New York

The Teeny-Marriage and Adultery

According to separation court records in the Archdiocese of Chicago, adultery was blamed as the second most frequent cause for broken marriages. Adultery was named in nearly one out of every four (24.8 percent) cases.

Just on my block there are three or four divorce cases and all because he or she was involved with another man or woman before they were separated.

Boy, 18, Milwaukee

The United States is the only major country in the world without a centralized system for collecting marriage and divorce records which would help keep watch on how its families form and break up. In the absence of true national reports, Americans are alternately jolted and soothed by published divorce rate figures which are bound to be inaccurate because they are incomplete. Adultery statistics are even more difficult to interpret than available divorce statistics.

American Catholic Family says:

AGE: Age at the time of marriage was of particular significance. When the husband was unfaithful, which was true in four out of every five cases, the highest relative percentages occurred where he was less than twenty-five years old at the time of marriage. When the wife was guilty, the rate of transgression decreased in direct relation to her age at the time of marriage.

NATIONALITY: The incidence of infidelity was relatively low among Polish males, Italians of both sexes, and Irish families.

INCOME: Divided according to occupation, members of the higher income groups engaged much more frequently in extramarital sexual contacts than lower wage earners.

Some marriage theorists say extramarital sexual relationships are inevitable because modern married couples have complex sexual needs arising from the divided, separate roles men and women must play at home and at work. That these needs are not being met by monogamy is a reality, they say, dramatized by the high U.S. divorce rate: four hundred thousand each year. The rate is not *a million a year* only because the cost of divorce and the cost of supporting a second family is intolerable. The choice for people who can't afford divorce is either desertion (one hundred thousand a year), or frustration, or an illegal love affair.

Marital fidelity is so spangled with rewards it is difficult to understand why anyone would ever choose the ugly troubles of infidelity. The furtive extramarital affair, by comparison, is a Rube Goldberg contrivance of time schedules and meeting places; it is impractical, inconvenient, undignified and expensive.

The way of the adulterer is hedged with thorns; full of fears and jealousies, burning desires and impatient waitings, tediousness of delay and sufferance of affronts, and amazements of discovery.

Jeremy Taylor 1650

Yet according to popular belief, very many married people do actually submit to these troubles by involving themselves in extramarital sexual adventures. One of Dr. Alfred Kinsey's reports is frequently quoted as the nucleus of this belief: "Close to 50 percent of our men and 26 percent of our women have had one or more extramarital relationships."

The trouble with this statistic is that it is used as if all acts of infidelity were identical. If we take the Kinsey finding at face value, we have to believe that most Americans suffer from a serious health disturbance, which is absurd. Psychiatrists are careful to make a distinction between *chronic* and *occasional* infidelity.

OCCASIONAL infidelity refers to the random, unplanned and almost accidental unfaithful act.

CHRONIC infidelity is adulterous behavior which is deliberate, frequently repeated and which follows a regular, predictable pattern.

The classic psychiatric stance toward chronic infidelity is that it is always a symptom of mental or emotional disorder. *Chronic* infidelity usually indicates a destructive sickness even if one does not physically commit any unfaithful act, but merely wishes or dreams it, as in the case of *psychic* infidelity.

The Chicago records show that men and women who married young got restless after a few years of marriage. They regretted being "tied down." They felt they were "missing something." Some wives had let themselves get too busy with the house and children and had neglected their husbands.

> Many girls get married and just don't know about the facts. Or if she had sex relations with someone before marriage and after marriage the husband finds out, there goes the marriage.
>
> *Boy, 18, Denver*

> Too many kids think marriage will just be one big sex party. Maybe the girl discovers that the boy is only interested in sex, where she is interested in what it should produce—growing love and children.
>
> *Girl, 17, St. Louis*

Five Alleged "Causes" for Adultery

Here, in no special order, are five "causes" married persons frequently allege to marriage counselors.

1. *Accidental meeting* of two persons sexually attracted to each other. These appear to be instances of *random* infidelity and usually point to little more than human weakness at a given moment of tension.

2. *Proximity at work.* In our times, fourteen and one half million women leave their homes and husbands each day and go to work in the presence of other men where they are exposed to flirtations, confidences and shows of affection. Their husbands, meanwhile, are working in the company of other women, some married, some single, some divorced—there are fifteen million divorcees in the United States. Travel to and from work adds to the length of time working couples are separated and adds further outside sex attractions. After-work cocktail parties, long business lunches and company get-togethers do their part to foster illicit office romances.

Many men and women who work together have more time and opportunity to communicate with each other than they have with their own wives and husbands. Sometimes they discover their work partners are actually more interesting than their home partner. In this light, it is rather a wonder that a majority of married people separated from each other by different work environments do, nevertheless, manage to maintain a spirit of nearness. They may be tempted to experiment outside of their marriages, but they usually reject the temptation.

Proximity at work or after work can be no more than a *circumstance possibly affecting extra-marital sexual contacts.* It cannot *cause* adultery. It can only make things easy for persons who are already looking for it. The marriage counselor will almost always find deeper intra-personal reasons at fault in infidelity.

Books, the cinema and theater, television, nightclub shows and advertising art are often popularly blamed for bad marital morals. These may possibly have some accidental influence in the forming of marital ideals, good or bad. They have a lot to do with creating the image we have of marriage and family. They are frequently the direct source of fantasies involved in certain forms of psychic infidelity. But in no ways can they be blamed as an essential cause of physical, chronic infidelity.

3. *Sexual incompatibility* or *sexual frustration* will be discussed more fully in a later section where they are alleged as a divorce factor. Do they lead to adultery? Yes. Necessarily and with all persons? No.

The so-called "sex drive" is a powerful force. If it is blocked in one direction, it will seek its way out in another.

4. "I was alone so often and I got lonely." The lonely person whose partner is away frequently or away for long periods of time admittedly has a real problem. Usually, the seriousness of the problem depends on *how* the partner is away. There are men and women who are home every day and yet convey the feeling of being away, and there are men and women who are forced to be away and yet, even in absence, are present. Every individual will suffer from feelings of loneliness, but pernicious loneliness in marriage is caused by communication collapse.

5. "I got drunk. I didn't know what I was doing." Alcohol plays three distinct parts in the act of adultery: it may be (1) a *tool* and (2) an *occasion* before the act; it may be (3) an *eraser* after the act.

In Thomas's Chicago survey, alcohol and adultery were blamed for *more than half* of the six thousand divorces. Almost always the male was at fault. In nearly one out of every five cases the drinker associated with a woman of doubtful character causing the court to presume adulterous behavior.

Does alcohol by its nature inspire adultery? According to popular but wrong impression, yes. There's a New York bar with a sign which brags: "Our martinis make you see double and feel single."

The false belief that alcohol causes adultery is rooted in very ancient soil. Countless stories, beginning with the Book of Genesis, and countless plays and songs since Genesis have linked alcohol with illicit sexual relations.

Some early thinkers taught that man's Reason, ruling from his head, was ever in danger of being assaulted by Animal Nature, which paced about his lower body. Alcohol opened the rickety gate separating these two powers, whereupon Animal Nature sprang up, overwhelmed Reason and then ruled man. The philosophers who taught this ridiculous image of man have long ago been discredited, but the idea that Demon Rum conspires with Evil Nature to defeat man still lingers among some people today.

Alcohol as a *tool* has an uncounted number of functions. The seducer swings it as a blackjack, the fearful or nervous swallow it to steady themselves. It drowns bad memories, squashes doubts. Some working men feel they have a special right to a Friday or Saturday night drunk. It's their reward to themselves for a week of work.

Mixed male-female drinking is probably the commonest occasion leading to adultery. Cocktail parties make accidental adultery easier since they give men and women a chance to meet in a non-critical frame of mind.

The repeated connection of sex and alcohol does of itself create for some, an erotic stimulus. Some men will admit they become mildly excited by the simple fact that a girl will accept an invitation to cocktails. There is often the hope far back in their minds that liquor is the first step toward something more intimate.

But bartenders and psychiatrists both say that the idea of adultery is almost always born before a bottle is ever opened. Men whose normal work puts them in touch with either alcohol or adultery or with both deny there is any strong bridge between the two.

Adultery is rarely a simple act. Usually it is served by an elaborate staff of deceits and lies, some faint, some gross. The reason for this is that *fidelity is built slowly*. Often it is involved in a network of family and community relationships. If these positive relationships are firmly recorded in the brain, they won't easily be uprooted.

Few healthy, intelligent people who want a good marriage will ever take a sudden, bold dive into adultery no matter how much they might drink. There are too many "Caution" "Slow" and "Stop" signs printed in their brains for the occasional "Go" sign to have much effect. What often happens is that a person will deliberately loosen one or more of his fidelities or trusts to give himself room. He may use alcohol to help him go through with the untying and loosening.

Dr. Frank Caprio in his book MARITAL INFIDELITY, says it is usually "the urge to commit an infidelity that leads to the drinking rather than the other way around."

The Inside Causes of Infidelity

There is a popular tendency today to say adultery is a mere "symptom of sexual frustration." Thomas rejects this. Rather he says, we must face the simple fact that "some individuals are unstable and easily tempted; others either because of past patterns of conduct or present circumstances, find monogamy both trying and monotonous."

Many men get tired of their wives or find them dull.

Girl, 18, New York

One gets tired of the others body.

Boy, 17, St. Louis

The immediate causes of infidelity must be interior, *inside* the unfaithful person, since this is where fidelity is.

There are many forces besides the simple, inordinate lust for sexual pleasure that can attract a married person into sexual contacts outside of marriage. There is "homosexual panic," by which a person seeks to disprove or mask the stirrings of deeply buried homosexuality through a great number and variety of athletic heterosexual performances. Psychiatrists believe the fabled lover, Don Juan, was a latent homosexual.

There is fear of impotence or frigidity by which a person seeks multiple sex contacts to assure himself he is capable of normal sexuality. Some men and women marry without ever truly committing themselves squarely to all the rules of marriage. A woman may deliberately marry a man she knows is cruel or sexually weak. His misconduct or inadequacy, she feels, will give her the right later to accept extramarital consolations. A man may marry a beautiful girl he knows is sexually cold. Back in his mind he may not want to give up his bachelor liberties. A woman may rationalize chronic infidelity with the admission that she married for economic security. She does not want to forfeit that security but neither does she want to deny her need for love.

There is general psychologic agreement that both fidelity and infidelity are traceable to the primitive mental and emotional habits formed during the first years of life and re-formed in early dating patterns.

Both the quality and quantity of a child's relationship with his parents have much to do with determining a pattern of marital fidelity. Of prime importance is the child's emotional security. This will go far in permitting him to clarify and affirm identity as a male or female. A little girl who is made to feel by her parents that it is really a wonderful thing to be a little girl will develop a respect for femininity. If she is sure of approval in these beginning years, she will later be able to exercise her female rights and duties without exaggeration or sense of inadequacy.

On the other hand, she may develop an excessive attachment for her father. If this is not resolved, as an adult she may marry a man who reminds her of her father, then be unable to make love to him. She will go to bed with men outside her marriage with whom she can sexually respond.

A little boy who sees himself respected, liked and appreciated as a boy and who sees his father similarly respected and appreciated for being a man will later in life accept the duties and pleasures of manhood without need to refrain from sex or indulge in it excessively and extramaritally to prove himself. A boy whose father is very passive and whose mother is domineering may find it exceedingly difficult to imitate or be like his father. Rather, he will develop feminine patterns in imitation of his mother. Later, as an adult, this may bother him. He may feel the need to prove he is a man by engaging in feats of sexual endurance.

The Pay Off

A discovered or confused act of infidelity will usually strike a marriage as a tragic crisis. Undiscovered it usually lies hidden in the memory as a regrettable, worrisome wrong, grown over only after a long period of time.

The minimum effect of infidelity, when it occurs to a dialogic marriage, is to cause a run in the texture of the relationship. It may go so far as to destroy dialog utterly. The precise evil of infidelity, here, is that it attacks the *foundation of trust necessary for dialog.* The guilty partner suffers a terrible dislocation after his betrayal of trust. Where there is a sensitive psychic attachment between the partners, he* may worry that his infidelity is somehow evident in the face or voice. He feels compelled to *act* or *pretend* in order to hide his wrong, but this, too, worries him for acting and pretenses are foreign to the dialogic relationship. He is disturbed by an intangible, invisible film which now separates him from his partner's presence. He wants to confess in order to restore his *self* to its former status of trustworthiness. But confession might excite anger and jealousy in the innocent partner, might inflict intense suffering, induce active mistrust. He is now thoroughly tormented by his dilemma. He cannot reveal the true reason for this torment to his partner as he would like to, so he must disguise it with a lie, an ignoble excuse, and so further complicate his mental state.

* WE USE "HE" BECAUSE IT'S TOO AWKWARD TO REPEAT BOTH FEMININE AND MASCULINE PRONOUNS THROUGHOUT. MARRIAGE COUNSELORS REPORT THAT FEMALE INFIDELITY IS AS COMMON TODAY AS MALE INFIDELITY.

Adultery and the Law

Through the years, adultery has been regarded as a moral evil, a grave crime against God, society and the family. In times long past, the adulterer was killed or tortured or, at the very least, humiliated.

A history of laws pertaining to adultery is several histories together. It is a history of *family structure.* It is also a history of *women's rights.* Laws reflect changes in *public* and *private attitudes* toward adultery. Underlying these changes are *modifications* in the *concept* of adultery.

The earliest known command against adultery appears in the Code of Hammurabi, King of Babylon (c. 2250 B.C.):

> If a man be found lying with another man's wife, they shall both of them die, both the man that lay with the woman, and the woman.

We doubt whether either of these harsh laws was frequently and literally applied. Hammurabi specifies that if the husband is willing to forgive his wife, the king might intervene on behalf of the paramour. According to Talmudic rules of evidence, two proper witnesses must have seen the full act and cried out a warning to the guilty persons reminding them of the consequences before the death sentence could be imposed.

Caesar Augustus was the first ruler to declare that adultery threatened the state. Seriously worried by the corrupting Roman family, Augustus enacted the *Lex Julia de Adulteriis* by which he denied that any husband had the right to kill his wife. If she was guilty of adultery, he must properly accuse her in a court. If he failed, her father was obliged to accuse her, otherwise any right-minded citizen must accuse her. If she was convicted, the citizen would be awarded a part of her property and dowry. Since adultery injured the state, the state would confiscate half of her dowry and a third of her property.

Justinian I, during his reign (527-565 A.D.) over the Byzantine Empire, ordered the entire body of Roman law to be complied and annotated. The *Justinian Code* subsequently published became the basis of law for much of the Western world. By this Code, adultery was a crime because it attacked the life of the patriarchal family, and therefore also the state. The penalty was banishment to a secluded place for a minimum of two years. The husband who brought his wife back into his home before this time was guilty as an accomplice in her debauch.

The double standard of sexual guilt is by this time clearly established in law. It originates in the very nature of the patriarchal family.

> They have required from the women a degree of reserve and continency which they have not exacted from the men, because in women a violation of chastity supposes a renunciation of all virtue; because women, by violating the laws of matrimony, quit the condition of natural dependence . . . because the children of the wife born in adultery necessarily belong and are an expense to the husband, while the children produced by the adultery of the husband are not the wife's, nor are they an expense of the wife's.
>
> Montesquieu, *The Spirit of Law*

The Oriental and Indian family, too, were strongly patriarchal. The dominant function of the woman in these lands was to help her husband extend his family name by giving him sons of his own flesh and blood. Adultery threatened the family line with falsification. The adulterous wife, in failing her husband's family, failed in her basic function in life. In China and Japan, she was expected to take her own life. In Korea, she lost her status as wife and became a slave.

This is not to say that the Eastern man could do no wrong. A man could have sexual relations with another woman without sinning against his wife, but he was not thereby licensed to spoil another man's family line. Neither could he have relations with a virgin for he injured her and her family by making her unmarriageable, and he injured some yet unknown and undeclared man who might have wanted to marry her and perpetuate his family line with her special help. He could, however, have sexual relations with a concubine without feeling unfaithful to his wife.

In many countries, the husband was allowed to punish an adulterous wife as he saw fit. Pope Alexander VII felt exasperated by this attitude. He reminded Catholics in 1665 that the "crime of passion" was really nothing but murder, that no husband had the right to "execute" his wife for adultery.

Little is left today in Europe of the ancient laws ruling on adultery. A survey of modern laws shows a tendency on the part of the state to withdraw from this matter. Sweden removed adultery from its penal code in 1942. In England's Matrimonial Causes Act of 1962 no punishment is named for adultery though it is ground for divorce. According to Italy's Civil Code of 1942, adultery of the wife is ground for legal separation, but adultery of the husband is *not* ground unless it constitutes a grave threat to the family; in any case, it is not a crime.

Adultery is still a crime in modern India, but the crime is committed against the innocent husband, who is therefore entitled to sue the adulterer in civil court. Japan struck adultery from its penal code during the Occupation by American Armed Forces in 1947, and Israel did the same in 1953. Adultery is a crime according to Germany's marriage laws of 1946, but the state will not prosecute on its own initiative as it would in other crimes.

Adultery is still a crime in France today and is heard in penal court. This court regards the wife who is unfaithful as potentially more dangerous and damaging to society because of her child-bearing function. It dictates that the husband commits a misdemeanor if his adultery is committed at home, but that the wife is guilty whether her action occurs at home or away from home.

In Iraq, the residue of the once mighty Babylonian Empire and the origin of the Code of Hammurabi, no remnant of the Code survives. Iraq's present law, codified in 1926, was influenced by English Common Law but is basically Islamic in origin. Adultery is a crime, but the accuser must produce four witnesses to the full act to prove it took place. Provided this condition is met, the ancient ruling of the Koran is applied: one hundred lashes for both guilty persons.

> If four witnesses are not produced, the accuser is lashed with eighty stripes and never again believed. The adulterer shall not marry save an adulteress or an idolatress and the adulteress none shall marry save an adulterer or an idolater. Vile women are for vile men and vile men for vile women.

Puritans in England and Switzerland put adulterers to death. Puritans in the American colonies were less bloodthirsty. There is a famous entry in the Maine Province records for 1651:

> It is ordered that Miss Batcheller for her adultery shall be branded with the letter A.

For all its fame, branding was rare in colonial Massachusetts. In some settlements, an adulterer was forced to wear a brilliant red letter "A" on otherwise somber garments. Puritan Elders could require persons proven guilty of adultery to write a confession which was read aloud to the congregation on Sunday. The congregation would vote on whether or not to excommunicate the member.

At the time of the Civil War, adultery was the second leading ground for divorce in the United States. Today it accounts for less than 3 percent of our divorces. Usually two proofs are required: evidence of sufficient *inclination* together with evidence of sufficient *possibility*.

The State has quit its former position as judge and punisher of adultery. It no longer regards adultery as a threat to national security.

In March, 1930, the Motion Picture Producers and Distributors of America published the following "Code" for the guidance of film makers: "Adultery, sometimes necessary plot material, must not be explicitly treated, or justified, or presented attractively." In 1959, the Supreme Court of the United States struck down this code in ruling that no state had the right to prevent the distribution and circulation of "Lady

Chatterly's Lover." State licensers of film exhibitors, who also act as film censors, regarded the Supreme Court verdict as applicable to films as well as books. It was feared our cinema would be flooded with films making adultery attractive but this hasn't happened.

Practically speaking, there is no external penalty today for adultery except divorce. We tend to think when a pattern of adultery repeats itself, it is a symptom of mental or emotional disturbance. The adulterer is referred to a doctor, priest or counselor, not to jail.

The change in attitude towards adultery marks the reduction in power and authority of the father and the gain in power and authority of the mother. It is not that the moon has eclipsed the sun. The sun still rules the sky but the moon is now visible in daylight.

Religions today continue to regard adultery as a serious moral wrong. Society, even fast, big city society, still finds it scandalous. If well-known persons are involved, the press treats adultery as sensational news. It is not interesting news if poor old Harriet and Harry commit adultery. Adultery is far from being accepted as normal behavior. It is still feared as a threat to family security and health. The press would not care to write it up if it were a common and accepted way of acting in our society.

For centuries, chastity was forced on women by laws which regarded them as intellectually and physically inferior to men and dependent on men for food, shelter and protection. These laws and the society they came from inflicted harsh penalties on women who disobeyed them. Even when laws did not exact a penalty, the woman's own nature often saw that she was punished. She suffered from fear of disgrace or of disease or both; she feared an unwanted pregnancy. Today, a woman can escape disease by medical treatment. She can prevent conception by taking pills.

But one penalty still hangs over her head. Though invisible, it is by far the worst penalty. It is the loss of trust and the loss of trustworthiness and the loneliness which comes with this loss.

If the punishment for infidelity today is a more refined, more interior punishment than in the past, marriage today requires a more refined, more interior, more balanced relationship than in other years. We use the word "infidelity" more frequently today than the word "adultery." Fidelity (Latin: *fides* = faith) was once the oath of fealty or allegiance a freeman made to his king or country. The slave did not take this oath; only the freeman had the power, right and dignity to make such a pledge. Infidelity, then, is a kind of treason of mind and body. There is no double standard when this term is used: the unfaithful man is as guilty as the unfaithful woman.

Ethelbert issued the first Anglo-Saxon Code in A.D. 600. By this Code, an adulterer had to pay a fine to the offended husband and buy him a new wife. Many prejudices and inequalities still exist, but the values set on women have been drastically revised since Ethelbert's day.

Why Fidelity?

FIDELITY IS THE SINCERE COMMITMENT OF OURSELVES, BODY AND MIND, TO ANOTHER WHOSE PRESENCE WE ARE AWARE OF EVEN IN ABSENCE.

With an increasing barrage of tests and opportunities daily attacking the traditional marriage, with current trends indicating a more benign public attitude toward extra-marital love, why do we continue to commit ourselves to the ancient, difficult ideal of fidelity?

Married people who are faithful to each other do not keep looking back to see how many days they have spent without being unfaithful—as if they had given up smoking. Marital fidelity is a form of health. People who enjoy it, like people who enjoy good, physical health, rarely take time to consider what it is. Our knowledge of fidelity, as a consequence, is weak. We pay much more attention to infidelity. Our interest is not necessarily morbid. We also know much more about disease and mental sickness than we know about physical and mental health.

But the neglect of fidelity as a subject of interest and study is not so much unfair as unfortunate. We are missing a fantastic source of strength. The ideal of fidelity has survived in spite of weak wills and strong temptations, in spite of negative examples from heroes and kings—worse, from parents. It has survived among all races in all continents from the first stone pages of history. Fidelity has not persisted because of customs or habit or religious doctrines. There is some deeper reason for its vitality. Here, with us, unexamined, is a power that has kept the family of man together, bickering murderously at times, divorcing, warring, but hopeful again and again, and moving constructively forward.

The sacrament of fidelity allows those who share it to develop a unique, exhilarating dialog, "a meeting of meanings." Sexual union is itself an intimate form of communication and is even called "conversation" and "intercourse." But the dialog we are referring to includes the sex act and is greater than the sex act alone.

Admittedly, there are people technically faithful to each other, who never achieve any exalted meeting of meanings. But where we *do* find such a marital dialog, we will almost certainly find that its foundation and its protective wall is mutual fidelity. Sexual fidelity in marriage is important because it germinates mutual honesty and trust. A dialog which has these two qualities will lead naturally to self-knowledge and therefore to strength. Such a dialog is the true food of personal development. It allows escape from the loneliness of self. It frees the persons involved from retarding mental and emotional narrowness and opens them to new experiences. It delivers the power to convert defeats into victories.

The example of a successful dialog is the best possible gift a married man and woman can give their children. Dr. Vincent A. Conigliaro, M.D., Assistant Professor of Psychiatry at Fordham University, describes the lasting value of this gift: "Children growing in a family where they can sense the mutual respect father and mother conceive for one another will quite naturally imitate this pattern and develop respect for it. On the other hand, children who see the father and mother arguing continuously or beating one another, verbally or otherwise, will feel that there isn't much to marriage and may show it later in extramarital escapades. The first image a child has of marriage is formed by what he sees in his own family. Both the quality and the quantity of his relationship with his parents will have much to do with determining his pattern of marital fidelity."

According to Mr. Frank Canny, a New York business and industrial psychologist, expert business administrators regard marital fidelity as an important company asset; they tend to give a second, suspicious look at the executive who chases secretaries or who worries about where his wife might be and what she is doing.

"You will read now and then," he notes, "of a small company owner dying and of his wife stepping behind the desk to continue the business. There must have been a close dialog between the two, else how would she know what to do?"

"Leadership is a lonely way of life," Mr. Canny explains. "Often the man at the top of a company depends only on his wife to stand by him when he's under stress and grappling with an extremely important decision. It's not important for her to understand the technical nature of his problems. He's just using her as a sounding board. When a man shows me a weak résumé of his career, I advise him to take it home to his wife. A wife is frequently the only critic a man will listen to. He would like approval for being brave or honest or clever. A man wants to win for somebody. If he loses, he likes to be assured by someone that he'll soon come back and win again. He can't easily hire anyone for this intimate, ego-supporting role. This is not a job for a mercenary. This calls for a complete, deeply committed loyalty. It calls not only for a good wife, but for a good marriage."

Married men and women living an open, guilt-free relationship, reflect a sturdy confidence that influences their work and affects their appearance and their communication with outsiders. A woman involved in an illicit love affair will resent housework because it's for the wrong man, or she will accept house drudgery as a secret penance to relieve her guilt. Either way, her health will eventually suffer.

Eileen Ford, whose famous New York agency books most of the high fashion models seen on television and in national magazines, told one of her girls who was having love troubles: "Nothing will break down a woman's beauty faster than a love affair outside her marriage, whether it's because she's afraid of losing her husband or because she herself is leading a guilty life. There isn't anything on the market that will disguise the loss of sleep and rest. The cosmetics inventor who discovers how to recapture the serenity of the happily married woman will become a millionaire overnight. There is no cream or powder in the world that can duplicate the kind of contented color and light that shines in a woman who knows she belongs to one man and that he belongs to her."

Promises

> Without being bound to the fulfillment of promises, we would never be able to keep our identities; we would be condemned to wander helplessly and without direction in the darkness of each man's lonely heart.
>
> Hannah Arendt in *The Human Condition*

The promise, like the resolution, is a kind of decision. When I promise to do a certain thing in the future, I commit a being, myself, who is in motion, growing and undergoing modifications even while speaking the commitment. I commit myself to another being who is also in motion.

A French philosopher, Gabriel Marcel, asks:

> How can I swear fidelity to act in any certain way toward someone ten years from now when I don't know who I'll be then? How do I know the person I bind myself to now will not alter seriously in years to come so that in ten years I can say: this is not the woman I committed myself to? Everything has changed, my mind and my feelings, her mind and her feelings. We are two vastly different persons, strangers from those we were ten years ago.

Marcel points out that because of change, *fidelity must always be renewed. Creative fidelity*, he says, is the ability of one changing person to keep trust with another changing person.

> No intelligent vow of fidelity can be made conscientiously unless it involves a promise of fidelity to God. Fidelity is an act of faith. All such acts are related to the final object of all faith. The self we pledge and the being we promise to are both in the image of a changeless God.

The simplest, most basic of all human needs is that need we first developed before we were born, while we were lying in our mother: the need to be *near* someone or for someone to be near.

> We need to be individuals but individuals in relationship. What we require is closeness, the absence of loneliness and isolation . . . Nearness provides emotional, social, spiritual and physical nourishment for the individual. Examining this need brings us close to the loves, hates and hurts of individuals both in and out of marriage.
>
> Dr. Aaron Rutledge

During early childhood, our need for nearness became more specific as it developed. We needed the nearness of certain persons we knew and liked. We needed to be cared for. We needed consistency.

The secret of fidelity's survival as an ideal lies with the role it plays in marriage: *to attend the human need for human nearness, and to provide for consistent love.* Fidelity is a power because it serves inter-human dialog. Only through the intimate man-wife relationship can man lose his harmful anxieties and affirm his true being. Dialog is the source of human *becoming.* It allows escape from the loneliness of self. It frees man from mental and emotional narrowness. It opens man to new experiences.

The ideal of fidelity won't be lost with the growing freedom of women because modern woman still needs a consistent personal love and presence. She can find this in a variety of relationships but in almost all instances, it is best provided by one man.

Chastity was once forced on women by the man's physical, economic and legal power over her. Today women are freed from these powers. They are equal to men before the law. They can work and can be economically independent. Contraceptive pills let her enjoy sexual intimacies without becoming pregnant.

Will the pill make her sexually promiscuous? No, it won't. If she wants to be promiscuous, the pill will help her escape certain penalties of promiscuity. The pill is not an aphrodisiac.

Persons who fear that easy-to-take oral contraceptives will encourage women to run wild reveal their own distrust of women and their failure to understand them. They may also be revealing their own secret lust for wild women! Of course, some women *will* run wild as some women always have. But even with freer sexual relationships possible, women still seek consistent, reliable relationships. Such relationships require intimate understanding, belief and trust. Fidelity is an outside act symbolizing these.

Infidelity corrodes the quality of nearness where it does not corrupt it. Lies, deceptions and disloyalties must be acted out in order to begin an illicit love affair. Lies, deceptions and disloyalties must be later constructed to hide or deny it. With each lie, the liar isolates himself a little more. Each lie is another clot interfering with the circulation of dialog and what eventually happens is that dialog dies. The final victory of infidelity is the breakdown of companionship and the return to morbid loneliness.

This is the power of fidelity: it allows the married man and woman to be and to grow. It clarifies their decisions. It supports and sharpens their dialog. It simplifies living. It helps each keep the promise to surrender self to the other. It is the fire which protects each from the menace of loneliness.

Quiz

1. Name three possible "outside" causes of adultery. Name three possible "inside" causes.
2. Precisely how does adultery affect dialog?
3. Why is sexual fidelity important in marriage?
4. Define fidelity.
5. What is the difference between *occasional* and *chronic* infidelity?

ENEMY OF BECOMING #3

Irresponsibility

You are obliged to be responsible from the minute you say "I do."

Girl, 18, Milwaukee

Couples who divorced because they were "irresponsible" made up Thomas's *third* largest category (12.4 percent).

But students ranked irresponsibility as the *first* cause of marriage breakdowns.

> I put irresponsibility first because if one or both of the parties has not yet grown up, the marriage has two strikes against it before it even starts. I feel this should be #1 because often the spouse doesn't realize that the other has not grown up until after the marriage ceremony.
>
> *Girl, 18, Milwaukee*

> Irresponsibility ranks at the top of my list and frequently couples rush into marriage before they can shoulder the hardships of married life.
>
> *Boy, 18, New York*

"Irresponsibility" is a rubbery category that can catch almost all marital grievances. Excessive drinking makes a person incapable of acting responsibly. Adultery is an irresponsible act. Wasting family money is irresponsible.

Students blamed irresponsibility on youth, immaturity and hasty marriage.

> People are stepping out of childhood right into matrimony.
>
> *Girl, 17, New York*

> These kids aren't ready for the responsibility they get. They still need someone to tell them what to do and how to do it and when.

They haven't matured enough yet to be a head of the house and be responsible for the lives and well-being of others.

Boy, 18, Milwaukee

Because of "rush" marriages, after awhile they get tired of this person they've been "shacked-up" with and go to someone a bit more handsome or prettier.

Boy, 18, St. Louis

But Thomas denies that youthfulness and inexperience are at fault. He says:

> A couple might well be young and inexperienced and even a little naive about some of the realities of life but still make an excellent adjustment in marriage as long as they were willing to accept responsibilities and attempted to adapt themselves to changing circumstances.

Irresponsibility in marriage means that the married persons avoid decisions related to their new state. They make the decisions of a single person, ignoring the fact they are now a couple or, perhaps, a family. They lie to each other. They defend their false self-images and refuse to correct the false images they have of each other.

Each keeps acting as if he were still single, as if he had not changed his goals and now had new needs. *He* won't pay bills, spends money as if it is his alone, won't find work or keep a job. He refuses to help with the house or children. *She* won't keep house. She uses their credit card as if the end of the month will never come. She sees nothing wrong with dating old boy friends. They leave their children unattended while they go to the movies. Their children run on the streets while they sleep late. They don't care who their children play with. Their children sleep in the car while they are bowling. These are forms of marital irresponsibility.

The individuals falling into the present category were not merely young or inexperienced; rather they displayed a peculiar selfish disregard for the rights of others. They did this, seemingly, not so much out of malice as out of mere intellectual and moral shallowness. They were children in the sense that they were never "socialized." They seem never to have learned that as social beings their actions affected others. They ignored the fact that as members of society they were responsible for the consequences of their actions.

John L. Thomas

ENEMY OF BECOMING #4

Conflict of Temperaments

Thomas's fourth largest category included almost one out of every eight cases (12.1 percent). Broken marriages in this category puzzled Thomas:

> The members of this group are not irresponsible and seem well adjusted in society. They do not commit actions which render them markedly odious to the married partner. The essential roles of support and sexual companionship seem to have been adequately fulfilled. Nevertheless, the partners find it impossible to get along together.

Both boy and girl students also made temperamental conflicts their fourth largest group. The students below feared such conflicts as the leading cause of marital breakdowns. They also seemed puzzled by the conflicts and offered various theories.

> Temperamental differences ranks first because today so much emphasis is placed on the individual's psychological make-up that people become preoccupied with being this or that "type" of person and not themselves. They are like children playing "grown-ups."
>
> *Girl, 18, New York*

> I think that many people fly off the handle too easily and this is a great cause of breakdowns.
>
> *Girl, 17, St. Louis*

> Because they're with each other constantly.
>
> *Boy, 18, Milwaukee*

> Most people are crazy in one way or the other and this seems to cause friction which causes divorce.
>
> *Boy, 17, Denver*

If the irresponsible person decides things in marriage as if he were still single and he alone mattered, the badly adjusted person *communicates* as if he were single and he alone mattered. He is too selfish to respect and consider anyone else.

Thomas was studying *dead* marriages. But marriage counselors who deal with family conflicts while they're hot in the air, say that most serious conflicts in marriage begin with communication blocks.

> Married persons, as the intensity of their love develops, tend to let go of their restraints, forget manners, tell painful and embarrassing truths. They get

brutally honest. Their frankness is stronger than their tact. They may feel quite safe to express hostility because nobody's going to run away.

As this interaction gets hotter, one or both persons goes too far and gets hurt or afraid. They cannot stand it. To make sure they won't hurt or get hurt again, they consciously or subconsciously put limits on their self-expression. They won't talk about certain subjects. Piece by piece other subjects are added. Gradually areas of life, thoughts and feelings are added and soon there is a collection of reactions, a no-man's land which must be avoided to avoid trouble and keep the peace. Thus communication within a marriage becomes blocked. Hurt feelings, misunderstandings, and resultant bitterness get bigger and collect until little except anger and rejection are being shared.

Aaron L. Rutledge,
Marriage Counseling: An Overview

ENEMY OF BECOMING #5

In-laws

In-laws are just people who are thrown upon one but one for "compatible" reasons must respect them all.

Boy, 17, Milwaukee

Thomas says: ". . . the problem of in-laws, like that of money, may lead to considerable tensions and may call for extensive adjustment by one or both spouses, but it is usually resolved short of marital breakdown." In Chicago, in-laws figured in less than 10 percent of the separations.

Just the same, "in-laws," as one student wrote, "can help a lot to wreck a marriage."

If the parents disapprove of the marriage to begin with, there will always be friction among the husband and wife.

Girl, 17, St. Louis

They're always trying to get their son or daughter on their side.

Boy, 17, Denver

2 women + 1 man = mess in 1 home.

Boy, 18, Denver

Immaturity is at the bottom of in-law trouble. The husband or wife isn't ready to leave home; father or mother can't let go.

> The first time something goes wrong, one has to go back to mother. Or mother forgets that her child has grown up.
>
> *Girl, 18, New York*

> I don't think that either of the marriage partners should spend more time with his parents than he does at his own home. If a person can't live away from his parents he shouldn't get married. I don't mean that parents should be tucked away in the back of a drawer or forgotten, or ignored, but they should be seen as they are. They should no longer be depended upon.
>
> *Girl, 17, St. Louis*

What kind of immaturity does in-law trouble involve? How does it show itself?

The *immaturity* is *evident* in *decision-making* and in *communication.* He's not used to making decisions that include her. She can't manage with the money he gives her. She calls her mother every day and they talk over an hour. He suspects her mother is advising her and his decision-making ego is offended.

> In the beginning of the marriage the couple must be getting used to one another—not mother and dad. Regardless how old a person gets, parents still feel pretty responsible for them and will try to step into marital problems nine times out of ten. The parents, though, don't really know both parties and cannot give a good opinion about how they should communicate. Whether they should use contraceptives or where the children go to school and church, but they will try anyway.
>
> *Girl, 17, St. Louis*

A wife and husband must develop their own system of communication and decision-making. They do this by trial and error, by mutual trust and encouragement. They can't easily develop this trust and can't freely experiment if critical outsiders are watching every move they make.

> In-laws should be told in no uncertain terms to stay out of your life.
>
> *Girl, 18, New York*

Some young marriages are spoiled by immature parents who resent losing the economic or emotional support of their son or daughter. The ultimate goal of parents is to make themselves superfluous. They must virtually *reject themselves.* Some are not mature enough to do this. They use tricks to make their child dependent again. They loan money. They pretend to be sick and need help. They insist their child is sick and bustle about playing nurse. They say they're lonely. They accuse their children of ingratitude and make them feel guilty.

Mothers-in-law are very good at ruining the children's home. They have run their household for years and would like to run their kids homes.

Girl, 18, Denver

Mothers and fathers can't realize that their children are now gone and on their own. They treat them as if they were still living at home and run their lives as they did when they were young.

Girl, 17, New York

The parents tend to have a strong influence if the married couple doesn't take a strong stand together.

Girl, 17, New York

I want my husband to love *me*; not my parents. Since *I* am going to marry him, I want *them* to "butt out." I expect his parents to do likewise. Our life will be *ours alone.*

Girl, 18, Milwaukee

The immature young mother won't let her child solve his own problems. She lifts him over obstacles, carries him whenever he cries. The immature parent will never let go of her child but will follow him right into his marriage and fight his wife over him.

ENEMY OF BECOMING #6

Sexual Incompatibility

"Sexual Incompatibility" is often a vague, catchall excuse used when the real origins of a breakdown are unknown or too embarrassing to talk about. The awkward, stilted language of students speaking about this subject shows we're not used to writing about sex. Many adults, doctors included, have the same stiffness of expression.

Because they can't give the marriage act to the other the other party wants someone who does.

Boy, 17, Milwaukee

By sexual incompatibility we mean the refusal or inability of married persons to have satisfactory sexual intercourse. The inability may be physical or psychological.

The man may be unable to achieve or maintain erection. He may ejaculate too soon or not at all. The woman may feel no interest or enjoyment in the act and end it unsatisfied by orgasm. Not all sexual contacts are the same and not all, by far, are perfect. It is only when married persons fail to satisfy each other for a long time and *fail to adjust to their failure* that we say they are incompatible.

Physical inability to have intercourse.

Sexual inability may be due to extreme exhaustion or advanced alcoholism, but repeated failure over a long period of time is rarely due to a physical cause or to malfunction of the sex organs themselves. *Sexual inability usually has a psychological cause.*

Psychological inability to have intercourse.

Either wife or husband may reject intercourse because of boredom or a new interest in someone else. The rejection may be *deliberate* or *conscious* or it may be *unconscious.*

For every psycho-sexual block, preceding it is a communication block. It is the communication block which is the enemy of becoming. Sexual incompatibility is the wound made by communication failure.

From Mr. T.'s earliest sexual engagements with his wife, he was able to maintain an erection only by imagining he was with a prostitute whom he had regularly seen years before when he was eighteen. At times he felt qualms from this deception, but decided it was harmless so long as his wife did not know. However, after five years of marriage, his wife gave birth to a lovely baby girl. Thereafter, Mr. T. suffered such guilt and distaste from associating or confusing the baby's mother in his imagination with a prostitute that he felt compelled to discontinue the fantasy. But this left him unable to consummate coitus.

Dr. Aaron L. Rutledge

Thomas concedes failure to achieve a good sexual adjustment can and does lead to a lot of frustration in marriage, but denies it is a *direct cause* for more than 5 percent of the breakdowns he studied.

High school girls ranked "sexual incompatibility" as a marriage's *fifth* most serious threat. Boys ranked it sixth.

A few students showed wise insight in their comments:

. . . I consider it more serious than religion because although there may be disagreement on religion there is less of a chance that the husband or wife will say something cruel or embarrassing in front of others as they might for sexual incompatibility. *Boy, 18, Denver*

But many student papers revealed misconceptions and prejudices about love, marriage, intercourse and man-woman differences. Is there anything wrong with the following comments?

Usually the man is more inclined to physical emotion and if he turns out to have a "frigid" wife, he might seek elsewhere a means to satisfy his passion. *Boy, 18, Milwaukee*

A woman naturally tends to be "colder" than a man—in either case if one or the other tends to be cold toward their spouse a complex may develop and all sorts of ideas come to the mind about love and how to show it.

Girl, 18, St. Louis

Man gets married because he wishes to enjoy the sex act. If he can't enjoy it, he can't enjoy marriage.

Boy, 18, New York

Should know better—try it before you buy.

Boy, 18, New York

Intercourse is not absolutely needed to make a marriage work.

Boy, 18, Milwaukee

Love diminishes when the two seem never to reach climax at same time—usually the wife.

Boy, 18, Milwaukee

A number of students confused "sexual incompatibility" with *sterility*:

I believe that if a couple is not able to have a child there is bound to be a break up in the marriage, for it is children that keep the family together. Even if a couple adopts they will still feel that its someone else's child not their own even though they do not admit it.

Girl, 17, Milwaukee

Rejection of Intercourse

A wife may accept her husband inertly and not be a true partner in love making. She may *pretend* to enjoy her husband and pretend to be sexually satisfied.

Why does a wife reject intercourse? For many reasons, but frequently because she is rejecting her husband as a person.

If she'd given me the proper sex I wanted, at least treated me like a person and not degraded me all those times, I wouldn't be going out to find out if I was a man or not.

In 1955 Judy was born when we were at Fort Dix. My wife called me to her bed. She had what they call natural childbirth. "The pain I went through," she told me. "Al, I'll kill myself if I have another baby. Promise me, Al, no more babies." I promised her.

Six weeks pass and I notice one of Judy's legs is shorter than the other, and her legs wouldn't open. Then we brought a doctor in. He says, "Your daughter will never walk again in her life."

There was no more for me. It was always Judy, always Judy, and this went on for one year, then two years. There was nothing there for me. So I cut out.

I asked the doctor. "She's so frigid," I said. But she was afraid to have sex because we might have babies. She said, "Our next baby might be born without arms"—like what happened to one of her girl friends.

She said to me, "If we're going to have sex, I'll let you know." *She'll* let *me* know! I used to think, what's wrong with me? Am I undersexed or oversexed or what? I bought some Kinsey books and read them. I wanted her to read them. She said "I don't want to read that kind of stuff." I said, "Well, let's go to the doctor, let's talk this over." But she don't want to hear nothing.

Albert DeSalvo's confession
from Gerold Frank's account
of *The Boston Strangler.*

A wife may reject intercourse to hurt her husband, as can be seen in this case described by Dr. Rutledge:

Sarah H. complained to the counselor that her husband Tim H. didn't talk enough with her. He would come home from work and watch the news. After supper, he was back in front of television again until bedtime. He wanted intercourse immediately, then fell asleep while she smoked cigarettes in the dark, angry at being used as a *thing*. She resented marriage as legal rape and began to loathe her husband's presence in bed with her. She yielded to him for a time, let him satisfy himself on her while she remained bitterly unsatisfied. But eventually, she pushed him away and had the bed alone. He slept on the couch. Her rejection became his excuse for seeking sexual contacts outside of marriage and led finally to their divorce. Her own sexual dissatisfaction "excused" her to do the same.

Some students—boys—excused men for avoiding intercourse because they had worked all day—as if women didn't.

Women expect a lot of love from the men and after a hard days work well some are too tired.

Boy, 18, Milwaukee

Sometimes a husband or wife seeks the love of another to make up for the lack of love at home due to fatigue or being busy with other matters.

Boy, 18, Milwaukee

It is true fatigue, worry, tension often make sexual intercourse difficult. But the incompatibility described in the case of Sarah and Tim was not due to any of these. The incompatibility was one of *communication*, as Dr. Rutledge shows:

> Sarah and Tim had never had a talking relationship. They lived separate lives and met only in bed. Their early communication was limited to *touch*. Tim ignored Sarah as a full companion and used her body for his physical sexual gratification. Home for him was a place to come to after work. It was a house full of ITs—television, supper, sex.
>
> Sarah, too, avoided talk till it was too late. She could never bring herself to tell him what he was doing wrong. She would tell the counselor: "Can't he *see* I'm unhappy? Is he *blind*? *I'm* not going to tell him. If he doesn't love me enough to be able to see that for himself I'm not going to bother."

Dr. Rutledge describes another example of rejected intercourse:

> Mr. and Mrs. K., Catholics, Detroit, have two children and don't want more. "Rhythm" as a reliable method of preventing conception hasn't worked for Mrs. K. The Church (to date, Fall of 1967) has not declared its practical norms toward the use of pills but Mrs. K's pastor has already told her she may not interfere with conception by using any chemical product.

The number of children may vary and the reasons for not wanting children vary too, but Mrs. K. is facing the dilemma of conscience familiar to marriage counselors as "the Catholic problem."

The dilemma is this: to use the pill, Mrs. K. must disregard her pastor's instructions or deny they apply to her. Her only other means of avoiding children will be to keep away from her husband and have him keep away from her.

Under the dilemma is a second problem. A marriage decision made unilaterally by one person and imposed on the other is only *half* a marriage decision. A decision which affects both members of the marriage should be made jointly. If the decision is one which concerns a matter of conscience, it should involve both consciences. A 'half' marriage decision cheats the family unit of dialog and becoming.

A husband who spends the rent money in a bar, for instance, is making a 'half decision.' He isn't spending his own money even though he may have worked to get it.

Similarly, a wife who decides secretly and alone to have a baby or to have an abortion or to prevent conception is making a "half decision." If she doesn't discuss the decision with her husband, even decisions which are matters of conscience, she steals a decision from him. She robs him and robs their marriage of becoming. If having a child is a cooperative act, *not* having a child should also be a cooperative act.

This book is neither a sex manual nor a guide to sexual morality. We will not attempt to solve Mrs. K's conscience for her. The question here is whether refusal to have intercourse leads to alienation and divorce. The answer to that question is *yes, it does*. Does it always and necessarily lead to divorce? *No*, it does not.

Thomas found about 5 percent of the Catholics in his study divorced because of sexual incompatibility. There is no further breakdown of this statistic. We don't know how many Catholics suffered psychologic or physical inability, nor how many husbands divorced because their wives refused to have intercourse with them.

But Thomas studied only marriage breakdowns handled by the Archdiocese. His Catholics still accepted the Church's authority. There is no record of the many Catholics who obtained a civil divorce and illogically and deplorably solved this agonizing sexual dilemma by abandoning religion.

ENEMY OF BECOMING #7

Mental Illness

Any mental or emotional disturbance handicaps *becoming* since it weakens or, if severe, paralyzes both decision-making and communication.

One student wrote, cruelly, pessimistically and inaccurately:

> No one can live with a nut.
>
> *Boy, 18, St. Louis*

Thomas assigned 3 percent of his marriage breakdowns to this category. He included all cases in which one of the persons had been put into an institution or judged mentally ill by qualified medical opinion. The reason given for separation was usually fear of violence or desire for security.

> I put mental illness first because if it cannot be cured the sick person is completely different from the one you married and you can no longer have a satisfactory relationship. So the person who is not ill would look for other companionship.
>
> *Girl, 17, Milwaukee*

What is broadly apparent in the student remarks about mental illness is that their ignorance of the subject has been enlightened only by mental health fund-raising slogans:

> It strikes every 30 seconds.
>
> *Boy, 17, Milwaukee*

> It is said 1 out of every 3 persons will spend some time in their life in a mental institution. The world has too much rush in it and not enough relaxing.
>
> *Boy, 17, St. Louis*

> Since 1 out of 4 Americans is mentally ill that leaves a lot of nuts around and if a person cannot cope with this problem, divorce is inevitable.
>
> *Boy, 18, New York*

> This person is angry all the time and doesn't want to do anything so the other leaves to seek someone more enjoyable.
>
> *Girl, 18, Milwaukee*

Some students correctly noted that what was really a manifestation of mental illness might appear as a more familiar disturbance. Sickness may show itself in an abnormal fear of sexual relations or of having children or of spending money or being home alone. It may reveal itself in deep, long and causeless moods of depression. Alcoholism, cruelty or severe and compulsive jealousy might also be forms of mental illness.

> Many broken homes are caused by mental illness which no one knew the person had.
>
> *Girl, 17, St. Louis*

> What really is mental illness people don't realize and so they blame their troubles on something earlier on the list.
>
> *Boy, 18, Milwaukee*

Divorce or separation does not in every case lead to freedom or mental health. Marriage has sometimes helped two emotionally or mentally ill persons to maintain themselves outside of an institution.

> Marriage is for better or worse. There are many new medicines and treatments so that the affected one can be cured with help and an understanding mate.
>
> *Girl, 18, St. Louis*

ENEMY OF BECOMING #8

Religious Differences

Most high school students ranked religious differences as the second weakest threat to marriage stability. Many appeared to think of mixed marriages as the only reason for conflict.

> Religion has been with a person from youth on and it's hard to change convictions.
>
> *Boy, 18, St. Louis*

> My mother a Catholic, resented my grandmother because my father, a fallen-away Lutheran refused to become Catholic while his mother was alive. Well, Grandma's still living, my mother still resents her, but Dad's a Catholic since two years ago Easter. I don't know how much this proves except that they got along ok in spite of religious differences. Religion was the least of their problems.
>
> *Girl, 17, Denver*

In Chicago, Thomas says, "Religion as a factor in the breakdown of marriage was surprisingly rare although seventeen percent of the cases involved mixed marriages."

One student offered this perceptive insight:

> I put religion far down in the list because "religion" is so rare today in people, by this "religion" I mean committed and dynamic faith.
>
> *Boy, 17, Milwaukee*

Whether religious differences are a weak or strong threat to marriage will depend on the convictions of the persons involved. Seldom are there fights over "Transubstantiation" or "Virgin Birth" even in mixed marriages. Troubles come from a conflict of consciences in deciding such matters as birth control and the education of children.

> Differences in religion – arguments about sexual relations, education of kids.
>
> *Boy, 17, New York*

Some married persons find it an agony to discuss these conflicts. A husband may refuse to talk about birth control and leave this decision entirely to his wife's conscience. On the other hand, some wives assume the right to decide how many children to bear is entirely their own.

> Religion is most serious. They don't keep the agreement they made before they married and then they bring up "Who's wearing the pants in this family?"
>
> *Girl, 18, New York*

But childbirth is achieved by a cooperative act. Child support and education are maintained by cooperative effort. So, too, is family-planning a cooperative act.

The husband who pushes birth planning decisions entirely over to his wife's conscience is running from responsibility and dodging both dialog and decision-making.

The wife who grabs up this decision as her sole responsibility is forcing her conscience on her husband. This is a serious violation of dialog.

The exercise of conscience in marriage should follow the principles for making good decisions. Agreement in exercising one's conscience is subject to the principles for dialog.

If there are strong differences of conscience regarding the birth and education of children they should be discovered and revealed *before* marriage during the dating period.

ENEMY OF BECOMING #9

Money Problems

In one way or another, money figures in most marriage problems. A husband may waste it on booze or a wife may throw it away on clothes. The lack of it may start fights or bring in the in-laws who start fights.

> Money always causes great mental strain—especially when there is none. Both become irritable and are still unable to cope with other problems of adjustment much less worry about bills.
>
> *Boy, 17, Denver*

> Sometimes the husband gambles or buys things for himself or the wife buys everything she sees. Or maybe they both have to work and just aren't making enough.
>
> *Boy, 18, Milwaukee*

> Wife blames husband for not enough money—she doesn't realize that the purpose of marriage is love come rain or shine.
>
> *Boy, 18, St. Louis*

Thomas blamed money for divorces only where disagreements over its use led to the break-up. He found this occurred in only one case out of every one hundred.

High school boys thought *money* was a bigger troublemaker than *sex* or *mental illness* or *in-laws.*

> Money isn't everything . . . but it's sure got a big lead.
>
> *Boy, 17, St. Louis*

Girls disagreed with boys over the seriousness of the money threat. They ranked money third from bottom as a cause for divorce.

> Money is a poor excuse for breaking up a marriage, if love is present and they pray to God or borrow until the debt is paid. Starving is better than breaking up a marriage. Of course, if children are involved it's a different story.
>
> *Girl, 17, St. Louis*

> Money is the last because small differences are usually easy to settle if they are settled right away and I think it is only normal that there be a few differences between two people living together. No two people are alike in taste and attitudes.
>
> *Girl, 17, New York*

Some boys don't date because they believe girls want them to spend a lot of money. These boys showed the same low opinion of married women.

> Women are brought up expecting a lot more and when the husband can't afford it, well . . .
>
> *Boy, 18, Milwaukee*

> Because many people feel that if they don't have it there is no use in living with that person and they go out and try to find someone that does have it.
>
> *Boy, 18, New York*

It's possible the boys developed this attitude from what they heard and saw in their own homes. Also, possibly, the complaint disguises B.S.I.—bad self-image. They feel they themselves can't attract or interest another person and therefore have to buy attention. They may have learned this attitude at home too.

If money means so very much to a girl that she would leave her husband for not providing enough of it, wouldn't this attitude show while they were dating? If it didn't show, wouldn't this fact indicate some horrible dialog failure?

The use of money in marriage is a decision-making problem. Agreement in its use is a communication problem.

Money decisions are directly related to a couple's image of a living standard. If this image is false or unrealistic or if they have extremely different images, there will be trouble.

> Money can be a big cause when both are used to having and handling their own. Differences of opinion on how to spend it happen very frequently simply because of the differences in nature of man and woman.
>
> *Girl, 18, New York*

Family economists like to blame money as a leading factor in divorce. They note that in 1963 almost 140,000 U.S. families declared bankruptcy. The trouble was not a lack of income but bad money and credit *decisions*.

If money is a marriage problem, it's because principles for decision-making have been violated.

THE PRINCIPLES	THE VIOLATION
1. Don't run from making a needed decision.	1. The payment of debts is postponed even when the money to pay is on hand.
2. Don't decide in an extreme mood.	2. "Impulse" buying frequently violates this mood. (Old joke: "Whenever my wife is down in the dumps she buys a new hat.")
3. Assemble, write down pertinent facts.	3. Failure to make out a reasonable budget. Failure to follow it. Failure to keep checking account balanced.
4. Separate major and minor goals.	4. This is the *key principle* involved where money problems are marriage problems. The budget should be organized to serve major goals. Failure to spend money according to these goals is what makes trouble.
5. Collect, grade advice.	5. It is easy to collect financial advice. There are bank advisors, insurance advisors, stock advisors. Salesmen, real estate agents and loan office managers are advisors. In-laws and friends are also ready to advise. Trouble comes when the advisor does not know or consider the couple's personal major and minor goals.
6. Act within your power to deliver.	6. Easy credit tempts persons to buy beyond their capacity to pay back.

A mid-western Family Service agency's survey found that *more than half* of the *family fights* it investigated were fights over *money* whereas less than a quarter were *in-law fights.*

A man and wife may make bad money decisions together. Both may confuse their major and minor goals. They won't quarrel because they have no communication problem. Communication problems start when either husband or wife violates one of the decision-making principles unilaterally—alone, secretively or without agreement or consent of the other. This act by itself is a violation of dialog. It is a decision *imposed* on the other person contrary to an agreement not to do so.

Marriage counselors find money disagreements are often a cover for deeper communication problems.

MARRIED 13 TIMES COUPLE PLANS TO WED IN 37 MORE STATES

St. JOSEPH, M. (AP)—Married to each other 13 times, Mr. and Mrs. James D. Grady plan 37 more weddings.

"We plan to be married in all 50 states of the union," Mr. Grady explained. "This is our way of protesting against divorce. We think being remarried is one way of maintaining a healthy respect for wedlock."

The Gradys overheard a newly married couple quarreling in Reno and that, they say, prompted their project.

Traveling in a mobile home, they hope to finish their matrimonial tour in the next 12 months.

License fees have ranged from $1.50 to $6 (they carry the licenses with them).

"It costs a little to get a license and pay the judge each time, but we're seeing a lot of the country," Mary Grady says. "Sometimes the judges think we are a couple of nuts, but most of them think it is a good idea."

New York Times
August, 1965

Quiz

1. What is marital irresponsibility?
2. How does irresponsibility affect dialog?
3. How does irresponsibility affect decision-making?
4. Can temperamental conflicts affect dialog? Explain.
5. Can temperamental conflicts affect decision-making? Explain.
6. Can in-laws affect dialog? Explain.
7. How can in-laws affect decision-making?
8. What does "sexual incompatibility" mean?
9. Can sexual incompatibility affect dialog? Explain.
10. Can sexual incompatibility affect decision-making? Explain.
11. Can religious differences affect dialog? Explain.
12. Can religious differences affect decision-making? Explain.
13. Can mental illness affect dialog? Explain.
14. Can mental illness affect decision-making? Explain.

CHAPTER SEVEN: THE MARRIAGE DECISION

Early Marriage

If the class using this book follows national statistical averages, *almost half of the girls present today in this classroom will be married before they are twenty years old.*

The chances of early marriage are greater in the United States than anywhere else in the world. So are the chances of early divorce.

Half of all American men are married before they are twenty-three years old. Only one-fourth of the men in most European countries and one-tenth of Irishmen and Norwegians are married by this age. One out of every six American girls between fifteen and nineteen years old is a wife compared to about one out of sixteen at this same age in France and Australia, and less than one out of fifty in Germany and Ireland.

In the United States, 94 percent of all the people marry at least once during their lifetime. A CBS television program titled "Is Marriage a Game for Kids?" said that two out of every five marriages in the United States involved at least one teenager. During 1966, 4,000 girls married at age fourteen and under. For several, it was their second marriage.

Unhappily, many of these young marriages fail. The Census Bureau reports that when both partners are under twenty-one, the divorce rate is three to four times as great as it is when they are between the ages twenty-one to twenty-five. In 1961, 70,000 teenagers in our population were either separated or divorced. Of this number, 22,000 were divorced and 48,000 separated. The teeny-marriage is a mini-marriage.

Of all marriages between teenagers, half will be divorced or separated by the time they are twenty-five years old. According to statisticians, the teeny marriage has only a fifty-fifty chance of lasting longer than five years.

Robert C. Cook, President of the Population Reference Bureau, makes this comment on our last census: "Today, more women marry in their eighteenth year than in any other; more have their first child in their nineteenth year than in any other. The thirty-eight-year-old grandmother will soon be commonplace!"

- One out of six teenage U.S. wives has two or more children.
- While only 3 percent of the 1959 baby crop had teenage fathers, 14 percent were born to teenage mothers.
- Of all babies born, the proportion born to teenage mothers was highest in the Southeast and Southwest and lowest in some Northeastern states, and Minnesota and Wisconsin.

The trend in early marriages is matched by a trend toward early dating. Boys and girls of twelve and thirteen are already pairing off as dates and going steady by fifteen and sixteen.

A letter to Abigail Wood's column in *Seventeen* under the heading "I've Been Dropped" begins: *"I've known Don since I was eight years old. We dated a few times early in high school but in my junior year we started going steady. After awhile, we realized we were in love and began making plans. We were to be married when I graduate . . . We even discussed the number of children we'd have . . .*

"Mark the approximate age at which you think teenagers should have their first date."

	1950	1962
a. Under 12	3%	1%
b. 13–14	44%	48%
c. 15–16	43%	46%
d. 16–17	5%	2%
e. Over 18	1%	0%
No response	4%	3%

Purdue Opinion Panel, 1962

Why Girls Leave Home to Marry

One way to keep from rushing the marriage decision is to go to college. According to Population Bureau, Inc., most women college graduates marry at age twenty-two; most high school graduates marry at age eighteen; most high school drop-outs marry at fourteen to sixteen years of age.

Some girls mature early, have simple, low-key ambitions and are content with little. They marry young but they stay married. Their lives aren't easy nor without suffering and even tragedy, but their resources of humor and patience help them cope with the hard turns their marriage course sometimes takes.

> I expect to do the washing, ironing, cleaning, cooking, love and share my life with my partner. I expect to get love and a home in exchange.
>
> *Girl, 17, St. Louis*

But many immature girls marry young without knowing what marriage is about. They say they realize marriage is a serious matter. They say they are marrying for life. But it seems obvious that they are not ready for more than the first three years of marriage. Why do they marry so young?

Often the girl is pregnant and has married for the sake of her baby or to avoid disgrace. Still more frequently, the teeny-bride is rebelling against her family. She wants freedom and independence. She wants to run her own life by her own standards. She doesn't want adults standing over her telling her what to do.

In 1955 a research project in Nebraska reported (1) that girls who married before age nineteen were emotionally less stable than those who married later; (2) that girls who married early often had unhappy relationships with their parents.

If a girl feels miserable at home, if she feels she's a Cinderella misunderstood and mistreated by her "wicked" parents, it's quite expected she'll marry the first "prince" who offers to "save" her.

The following tape-recorded statement was offered to St. Louis students for comment:

> "I was bored miserable at home. My mother nagged my every step around the house. My father was off somewhere in his mind. He didn't know I was alive. I made up my mind to marry the first guy who would take me. I figured marriage couldn't be worse than living at home."

The students' comments:

> Wrong. Unless the person was lucky enough to find a fine guy on the first try.
>
> *Girl, 18, St. Louis*

I understand. It's lousy living at home like that. But she could have maybe moved out, got a job first, given herself time to mature. If she wasn't of age to do this she should have maybe tried to see her counselor (altho they don't help too much).

Girl, 18, St. Louis

She used marriage as an escape. Actually it is an escape from nothing.

Girl, 17, St. Louis

This is definitely wrong. In fact, I know a girl who is planning to get married this July because she can't get along with her father at all. I've tried telling her and so has her aunt. I don't know how to help her, but I won't give up til she's gone for sure.

Girl, 17, St. Louis

This is the parents fault.

Girl, 18, St. Louis

This is terribly wrong. I know of a case like this where a woman got married only to escape crummy parents. It ended in heartache, constant arguing and divorce.

Boy, 18, St. Louis

Happened in many cases, might succeed because the nagged person would try harder.

Boy, 18, St. Louis

She wanted to prove to herself and mostly to her parents that someone else could give her what her parents didn't. And that is love.

Boy, 17, St. Louis

Go out and rent a room.

Boy, 18, St. Louis

Seems to be pretty common. At least by living at home she is bored only temporarily. Marriage is permanent.

Boy, 18, St. Louis

She will get married with an immature mind. Only two mature people can make a marriage last.

Boy, 18, St. Louis

It could just be a little worse when you come home after you break up.

Boy, 17, St. Louis

Wait until she sees the same man sleeping next to her thousands of mornings in a row.

Boy, 17, St. Louis

At one time, Tennessee tried to prevent spur-of-the-moment marriages by requiring the prospective bride and groom to post a $1,250 bond. Almost all states today require a waiting period or have other laws designed to stall easy "skid" marriages. Such laws do prevent many hasty and reckless marriages, but if two immature persons make up their immature minds to get married and are old enough and healthy enough to qualify for a license there is no law in any state that can stop them.

LEGAL AGE FOR MARRIAGE, BY STATE (AS OF AUGUST, 1958)

State	With Consent		Without Consent	
	Male	Female	Male	Female
Alabama	17	14	21	18
Arizona	18	16	21	18
Arkansas	18	16	21	18
California	18	16	21	18
Colorado	16	16	21	18
Connecticut	16	16	21	21
Delaware	18	16	21	18
District of Columbia	18	16	21	18
Florida	18	16	21	21
Georgia	17	14	21	18
Idaho	15	15	18	18
Illinois	18	16	21	18
Indiana	18	16	21	18
Iowa	16	14	21	18
Kansas	18	16	21	18
Kentucky	16	14	21	21
Louisiana	18	16	21	21
Maine	16	16	21	18
Maryland	18	16	21	18
Massachusetts	18	16	21	18
Michigan	18	16	18	18
Minnesota	18	16	21	18
Mississippi	17	15	21	21
Missouri	15	15	21	18
Montana	18	16	21	21
Nebraska	18	16	21	21
Nevada	18	16	21	18
New Hampshire	14	13	20	18
New Jersey	18	16	21	18
New Mexico	18	16	21	18
New York	16	16	21	18
North Carolina	16	14	18	18
North Dakota	18	15	21	18
Ohio	18	16	21	18
Oklahoma	18	15	21	18
Oregon	18	15	21	18
Pennsylvania	16	16	21	21
Rhode Island	18	16	21	21
South Carolina	16	14	18	18
South Dakota	18	15	21	21

Tennessee	16	16	21	21
Texas	16	14	21	18
Utah	16	14	21	18
Vermont	18	16	21	18
Virginia	18	16	21	21
Washington	14	15	21	18
West Virginia	18	16	21	21
Wisconsin	18	15	21	18
Wyoming	18	16	21	21

Paul H. Jacobson, *American Marriage and Divorce*, 1959, Rinehart & Co.

The Ideal Age for Marriage

Thirteen hundred and forty students in St. Louis, Milwaukee, New York and Denver were asked to circle the ideal age for marriage:

Woman: 16 17 18 19 20 21 22 23 24 25 26-29 30-33
____ other

Man: 17 18 19 20 21 22 23 24 25 26-29 30-33 34-37
____ other

A majority circled 20-22 for the woman and 22-25 for the man.

WOMAN'S	IDEAL AGE FOR MARRIAGE	MAN'S
1	**15**	**0**
4	**16**	**1**
10	**17**	**1**
21	**18**	**1**
100	**19**	**2**
320	**20**	**10**
470	**21**	**100**
270	**22**	**220**
100	**23**	**330**
40	**24**	**320**
20	**25**	**200**
14	**26-29**	**130**
0	**30-33**	**30**
0	**34-37**	**27**

The comments which follow suggest that a majority of students hold a double image of maturity, one kind for the boy, another for the girl. The students say a girl is mature when she is emotionally stable. A boy must be emotionally stable too but it's equally important that he be *financially* stable.

According to the students, the boy's maturity depends on a good job, money in the bank, or education—which relates to money earning. He must be emotionally secure in that he's had his fill of travel and of running around with girls and is ready to settle down and work for a family. The girl matures after leaving home by working for a year and meeting a variety of men. Only a small number of girls gave themselves time for their intellectual development.

WOMAN: 16 MAN: 17

The earlier the better.

Boy, 17, Denver

WOMAN: 18 MAN: 19

She is just out of high school and has made her decision on what she wants to do in life and the same is true for a boy and he has worked a year to save up some money for their future.

Girl, 17, Milwaukee

WOMAN: 19 MAN: 21-24

The most important consideration is that you marry at a safe age. An age when you feel that you will be able to make it. If a person has the maturity, money and a true love he could marry at any age.

Boy, 18, St. Louis

At 21 he is considered a man by the state, he can vote and he can drink. If he were much older he would be too far away (in years) for his children. 19 is ideal for a girl because a woman normally should be 1 or 5 years younger than the man because of maturity and because the man is head of the family.

Boy, 17, New York

At 19 the girl is at the peak of her life—she may love someone very much but is not able to express this love. There can be no set age for marriage. Different people mature at different ages and no one by themselves can determine this age.

Boy, 18, St. Louis

I plan on being married at 19 and my husband to be will be 21. I feel and he feels that we are capable of taking on this great responsibility.

Girl, 18, Milwaukee

At 19 a woman has had a chance to meet men and should have had the time to have the fun of a single woman.

Girl, 17, Milwaukee

At the age of 21, a man is legally of age and need not depend on his parents whenever he wants to buy something. By 19 most girls have tried their luck in the world in some form or other and are now ready to settle down.

Girl, 18, Denver

Woman should go to night school and work about 1 or 1½ years after high school and prepare for marriage.

Girl, 17, St. Louis

WOMAN: 20-21 MAN: 20-22

At this age the people are old enough to know what is happening and to realize if what they are doing is right or wrong. I myself am getting married in July and I'm only 18. The girl is 16.

Boy, 18, St. Louis

If you know the person long enough to know all his faults and good points and that you love him. By the time we're 20—we'll be going together for 6 years. I think I should know him well enough by then. And if I don't know him well enough I still love him and will marry him before I'm 21.

Girl, 18, Denver

WOMAN: 21 MAN: 23-25

Man has time to complete 4 years of college and one year to establish a job, a home, and a bank book. Woman has time for two years of junior college and two years to make money for clothes and some of the things she wants after marriage.

Boy, 18, Denver

By this time both parties will be able to cope with the difficulties that lie ahead. The boy will be out of college and will have the necessary credits for a decent family supporting job.

Boy, 18, New York

Woman is at legal age where she can have the privileges of an adult. 21 years should be plenty of time for a woman to be experienced and know what she wants in married life. Man can go through college and service and will need a companion.

Boy, 18, Milwaukee

24 is ideal for the man because he has already finished college and has been exposed to all kinds of women and can decide which one he would like to marry.

Boy, 17, Denver

21 and 24 are ideal, however, I must admit they are impractical because people usually can't wait that long.

Boy, 18, Milwaukee

Time enough for them to have enough money to be happy. Put a down payment on a house or have enough to pay the rent for several months. Man settled in his job has received a couple of raises and should know where he is going.

Boy, 18, New York

I do think people should wait to get married until they reach a reasonable age of about 21.

Girl, 17, New York

WOMAN: 22 MAN: 24-28

At 22 she is sure of the type of man she wants. At 24 a man has a B.S. and maybe a M.A. If not, he has had a good job for two years after college. If the people wait much longer they can become too set in their ways.

Boy, 18, St. Louis

You've had time to window shop and about these ages the merchandise is ripe for picking—you don't want your fruit too green or turned sour.

Boy, 18, Milwaukee

A girl who is 22 has had a chance to examine men and boys, thus seeing what life is all about.

Boy, 17, Denver

He must be about 25 years old . . . it takes time to save money. I don't mean he has to have a mint but I don't believe that two people can live on love either.

Girl, 18, New York

Because in most cases, the woman will be about as mature as she is going to get outside of marriage.

Boy, 18, St. Louis

Give a girl a chance to look around . . . attend some kind of school, etc . . . give her time to save around $1500, $2000 for marriage.

Boy, 18, Milwaukee

For a boy to marry before 26 is goofy. If married sooner he could not have had any fun in life with the service and school. By this time he should have a good nest egg for help.

Boy, 18, New York

Woman: she has some time to learn how to act in society, which will be beneficial for her husband. She will have had some kind of work probably to teach her money problems. Man: he has finished his education, had time to get a job, had experience with many girls, settled as to his wandering instincts.

Boy, 17, Denver

By 22 years old she has finished her schooling and has done most of her running around. She is emotionally mature enough to withstand the rigors of married life.

Girl, 18, New York

They should have some money saved to have a good start without being deep in debt.

Girl, 17, St. Louis

Any boy that gets married before he has completed his education and more or less started in his work needs his head examined. A girl can find satisfaction in the home but a man also needs his career.

Girl, 18, St. Louis

He should have a nice bank account and a knowledge of the value of a dollar.

Girl, 18, Denver

I don't think boys are grown up until they reach about 23 and then *I wonder.*

Girl, 17, Milwaukee

Because I am a woman I feel that before settling down one should go out into this world and experience both the ups and downs. Then when she reaches say twenty-two years, she can better appreciate the sacrifices her husband is making everyday for her and their children's welfare . . . Chances are that there will be no Cinderella story for most so it is better to be prepared.

Girl, 18, New York

That gives a girl four years of college to fall back on if ever her husband is sick or dies and she has to work.

Girl, 18, Denver

WOMAN: 23 MAN: 25-28

By 23 she will be a mature woman and not just a girl. She is definitely seeking marriage so that she knows what they are getting into.

Boy, 18, Milwaukee

I just think when 20 or under, married women are just infatuated and not in real love. A man should have completely matured and have had all of his liberties and freedoms enjoyed to the utmost. Their dowry should be big enough for a down payment on a house.

Girl, 18, St. Louis

WOMAN: 23 MAN: 25-28

If he was any older, he would be too set in his ways to compromise at all with a girl.

Girl, 18, Milwaukee

The woman will probably have made the rounds with other guys and will be able to choose her mate intelligently.

Girl, 18, St. Louis

By the time the woman is 23 and the man 25, both would be mature (or should be) and would have worked and maybe have money for a house and furniture. If you marry without these articles or without any money in the bank, you can have a happy marriage, but immaturity and financial problems can cause unnecessary problems in later life.

Girl, 17, New York

By this time they should each have a little money saved—enough to get them off to a good start so they won't have to borrow from either of their parents.

Girl, 18, St. Louis

At this age both parties have seen life in an entertaining way. They have also met problems and are better prepared for the hardships in marriage. Also you will most probably marry with a level head not a frustrated heart. You will be willing to share yourself if you have had all your excitement of the world before marriage.

Girl, 17, Denver

WOMAN: 24 MAN: 26-28

The man has had time to finish schooling, get a job, mature and become self sufficient and self reliant. When he would marry he would *not* be seeking a nursemaid or a "puppy love." By this time he would have been able to save enough money for the heavy expenses of marriage. The woman has had time to mature, become somewhat independent and self-reliant. By this time she would *not* be seeking a "hero" type or a father type. She would know the financial brutality of the world and could better help a mate. Both would have settled down—emotionally, sexually and financially.

Boy, 18, St. Louis

I don't think marriage while one is attending college is impossible only if you are still forming your basic attitudes about important things you aren't ready for marriage. Courses like philosophy and theology in college could change your thinking a lot. 20 (women) and 22 (men) is not too early if neither goes to school.

Girl, 18, New York

Get feet on the ground and rid of all urges for travel.

Girl, 17, Denver

I don't think either the man or woman is completely mature and free from the frivolous ways of young people until about this age. It is probably also better for them financially.

Girl, 18, St. Louis

WOMAN: 26-29 MAN: 29-33

They should live first, then marry.

Boy, 18, New York

Personally as a woman by 28 or 29 I feel I shall be emotionally and mentally mature enough to marry. Also because of the profession I have chosen I shall not be out of med. until about that age (and I don't believe in college and marriage mixing).

Girl, 18, New York

Ideal age for marriage is entirely dependent upon the character and degree of maturity of the persons involved. One standard to use though would be the amount of education of the couple. A complete college education is almost necessary for a man. The girl should also have as much education and cultural background as possible in order to be a better Christian wife and mother.

Girl, 17, St. Louis

Educate women as well as men for careers. Maternally allied professions should be held up early as careers ideal for girls. Even during their fifteen or twenty "mothering years," there are commercial and non-commercial activities suited for women's part-time or full-time assistance.

Lucius F. Cervantes, S.J.

Early Steady-Dating is the Way to a Teeny-Marriage

A famous family-relations expert believes that a good sex education course in high school has the effect of postponing marriage. Dr. Paul Popenoe, founder and director of the American Institute of Family Relations, says that students who take such courses delay steady-dating till they're older.

The dating period is a fact-gathering period. It's a time for getting information about self and others which can be used toward a good marriage decision, whether that decision is to marry now or later or not at all.

Steady-dating is an intense exposure to the facts.

"Are you now or have you ever gone steady?"

This is the question Purdue University asked about fifteen hundred high school sophomores, juniors and seniors in all parts of the United States:

Boys	Girls	
52%	61%	YES
43%	34%	NO

It's a waste of time to bluster and threaten that young steady-daters will certainly get into trouble because they may not. Even when it seems inevitable that steady-daters are going to bolt into an early marriage, we can't predict with certainty their marriage will be unhappy. Though the odds run heavily against successful teenage marriages, they may just beat the odds. Some of our own happily married parents steady-dated in high school.

Are there advantages to steady-dating? Yes, there are.

1. A "steady" may save wear and tear on the nerves. If a game or dance is coming up, it's good to be able to count on having a date.

2. Steady-dating may carry some prestige. This will depend on who your steady is and what your friends think of going steady.

3. Steady-daters have a chance to test themselves in many different moods, in many situations, with different problems. They can learn their weaknesses and strengths.

4. They can practice *complex* decisions which affect two persons, not just one decider alone.

5. Steady-dating can help develop good communication and dialog abilities.

Steady dating, these viewpoints considered, can be a maturing experience. But what are the disadvantages? What risks do steady-daters run? What do they lose?

The steady-dater, by investing time and attention on one person, loses the experience of *variety* in communication and decision-making. The steady-dater is not *free*.

> I was asked to a prom and my boyfriend wouldn't let me go. He wouldn't try to understand that it's the most important event in a girl's life. He wouldn't take me so I couldn't go. That was the final word.
>
> *Girl, 18, Denver*

> Ask your steady if it's alright with him if you went out a couple of times with someone else. If he says no, go out anyhow. You're not married. Just going steady.
>
> *Girl, 17, Milwaukee*

> You don't *stop* liking someone when you drop a steady. You just realize that there are an awful lot of other people in the world who you also like.
>
> *Girl, 18, Milwaukee*

> You're not sure enough of yourself to go steady.
>
> *Boy, 17, Denver*

> Every now and then I like to get out of steadying and get back in circulation. If we really had something we would eventually get back together.
>
> *Boy, 18, Milwaukee*

> You would like to have the chance to meet other boys, you feel you're being deprived.
>
> *Girl, 17, St. Louis*

> I'm not ready to settle down yet. I want to go out with other guys, lots of them.
>
> *Girl, 17, New York*

> Steady dating interferes in your personality development etc.
>
> *Boy, 18, Denver*

> You don't want to be tied down. You want to play the field. If you really liked the person its going to be hard not to see her all the time but it's better for both of you.
>
> *Boy, 18, Denver*

> It isn't fair for him or her to monopolize one another. They should give one another a chance to go out with different people and then come back together if they find out other ones are worse.
>
> *Boy, 17, St. Louis*

> You will never really know if he is the one if you are always with him.
>
> *Girl, 18, New York*

> I'm not the type that wants to be leached on to. You may say this in many different ways.
>
> *Girl, 17, Milwaukee*

> I don't want to get serious, and maybe into trouble. I want to be free.
>
> *Girl, 17, Denver*

You can't have a bad date with someone until you know both their good and bad points. It usually comes after going out with the same person too much, you lose that little something. You don't have as much fun.

Girl, 17, St. Louis

I saw too much of this one person, there was nothing left to talk about.

Girl, 18, Milwaukee

Steady-dating is a test of fidelity, that is, trust. At sixteen or seventeen or eighteen or nineteen are the daters ready for this test? Do they want it so early? There is no commitment in early casual dating, not even the commitment to date the person again. Commitments begin with regular, long-run dating. Steady dating tests the sense of commitment. Is the dater *ready* to *commit himself to one person so early in life*?

We don't say sexual intercourse and its serious complications are *necessarily* a part of steady-dating, nor is single-dating necessarily a sign of virginity. There are persons who have sexual relations during one-time-only dates, even during double dates. On the other hand there are steady-daters who don't even kiss.

But it is certainly easier to have sexual relations in a steady relationship than to have them during random dates or when not dating at all. It is natural for a man and woman who are alone together frequently for long periods of time to communicate more and more intimately in their language and by touch.

Steady-daters may be forced to make decisions they're financially and psychologically not ready to meet. They may find their desires and actions are in serious conflict with their moral and religious training. If they decide to go against this training, they must struggle with guilt.

Each of the questions below involves a serious problem of decision-making. Each requires skillful and mature communication between daters themselves and between the daters and their guardians and parents.

SHOULD I USE CONTRACEPTIVE PILLS OR DEVICES?

IF I HAVE A BABY, SHOULD I KEEP IT OR GIVE IT UP?

SHOULD I TRY TO GET AN ABORTION?

THE SCHOOL IS AGAINST ME FOR GOING STEADY—SHOULD I QUIT SCHOOL AND GO TO WORK?

MY FAMILY DOESN'T WANT ME TO GO STEADY—SHOULD I RUN AWAY?

SHOULD I GET MARRIED NOW?

Several boys said the only way to break out of a steady-dating pattern was to "avoid the girl." Suppose one of these boys should get a girl pregnant. Will he admit he's the father? Do you think he's ready for the responsibilities of fatherhood?

The emotionally involved steady-dater may be so fogged by feelings that he or she botches important school and personal decisions. If steady-dating disturbs the development of decision making skills, it handicaps the chief means for *personal becoming.*

Maturity

In our society "immature" is a popular word for putting someone down. As some persons use it, "mature" means behavior they like, while behavior they don't like is "immature."

What *is* maturity? The concept must be one of the oldest in civilization. Possibly, man learned it from the farmer and the farmer learned it when he discovered what ripeness was.

He learned, for example, that a grape vine took at least four years before it began to bear grapes properly. When this happened the vine was "mature." Then he learned that a vine got better and better with age and that in about its twentieth year the vine was at its best. He saw in this that ripeness had degrees. When the vine bore its best grapes, it was fully mature.

Something is mature when it achieves the goal expected for it or does what it's expected to do. In this sense, *maturing is becoming more.* The grape *becomes* its best possible self in autumn after a certain number of summer sun days. The vine needs about twenty years to *become* its full possible self.

Is a car "mature" when it runs? Is a clock "mature" for telling the right time? No. These are machines. We use "mature" only with *living, growing* things, things which *progress* and *change* as they do.

What are the signs of human maturity? How do we know when someone is mature enough to vote, see an "adults only" movie or read an "adult" book?

Human maturity progresses from a condition of total dependence to a condition of independence, from a condition of receiving everything to one of balanced giving-getting.

The ultimate, overall, kingsize complete and perfect human maturity is *sanctity*. It is knowing one's Self and *being* it while *becoming* according to the Self's potentiality.

Human beings mature *physically, psychologically* and *socially*, but not necessarily at the same time nor at the same rate of progress. Some persons develop intellectually but go backwards emotionally.

A young girl may be ignorant about sexual intercourse but be physically ready to become pregnant and bear a child. A young man may be able to explain articulately the meaning of the marriage contract, but not be mature enough to meet its demands.

Before an adult can assume he is emotionally mature, he must have settled most of his childhood and adolescent conflicts and be free to meet the conflicts which will be thrust on him by his age. An adolescent person is not mature if he meets the conflicts of adolescence with the same tricks and reactions which got him through childhood. An adult person is immature if he reacts to problems as he did during adolescence.

> To a certain extent, defenses established in early childhood remain unmodified by subsequent changes. To a certain extent behavior remains overly determined by parental attitudes valid in childhood but not in adult life. To a certain extent everyone deals with reality as if it were an echo of childhood rather than a new experience. If a person is to grow toward well-balanced maturity, he will find it helpful to have had secure childhood foundations. He should not have to divert his energies into fighting childhood conflicts and nursing old hurts.
>
> Alfred R. Joyce, M.D.

Our laws use a rule of thumb way to measure maturity and go by age. Six is the age for losing baby teeth. Seven is the "age of reason." Depending on your state, you may drive a car at fifteen or seventeen, but in all states you can't vote until you're twenty-one.

New York State and New Jersey are at bitter odds with one another over the legal drinking age. New York allows liquor to be sold to eighteen-year-olds. New Jersey says an eighteen-year-old is a minor and that liquor may not be sold to anyone under twenty-one. On weekend nights, New Jersey State Troopers wait at the bridges and tunnels to give balloon tests for sobriety to teenage drivers coming back from Manhattan. Their bag of arrests each week is heavy.

Reckoning maturity by age serves some purposes but it's a crude and arbitrary measure. More exact evidence of an individual person's maturity can be seen in his decisions and communications. These are the means by which he grows and *becomes more* as a person. A grape grows by means of sun, soil and rain. Given a season drenched with sun and no frosts, the viticulturist can predict a good harvest. Similarly, the human being's progress toward maturity is marked by good decisions and by dialog.

Is maturity achieved by nature's processes alone or can we do something to direct these processes? Can we speed them up or must we stand by and let them run their course?

A certain maturity does come to growing beings automatically with time. Wild grapes, for example, mature and become edible without anyone's help. But the viticulturist can enrich the grapevine's "experience." Helped by fertilizer and pruning, grapes can realize a fuller maturity.

We can help the progress of our own maturity (1) by wanting to be more mature, (2) by knowing more and more about our Actual and Potential Self, (3) by an open attitude towards difference and change, and, (4) by exercising the rules for decision-making and dialog.

Quiz

1. Explain: "Early steady-dating is the road to early marriage."
2. Describe five advantages gained from steady-dating.
3. Describe five disadvantages of steady-dating.
4. What is maturity?
5. We can help ourselves progress in maturity by four ways. What are the four ways?
6. What are the two chief means for marital growth?

The Marriage Decision

Marriage is not a game. One who enters marriage with the idea of an everlasting romance is due for a shock.

Boy, 18, Milwaukee

The decision to marry or not to marry, to marry this person or another, to marry now or later—will be the most important decision most of us will make in our lives. How will we attack this decision? Do we have a strategy?

We will address this decision as we do any important decision:

(1) Before deciding, we will get all our facts together. We'll write them down so we can keep them in front of us where we can organize them.

(2) We will get ourselves alone somewhere. Turn off TV. Pray for open, honest light. We don't decide if we're too tired to think straight, or if we're very worried or disturbed. Neither will we decide if we're out of our heads with excitement. Cool it. Take time. The marriage decision isn't a crash-landing decision. It has to be good enough to last a lifetime.

(3) We will separate our big, long term goals from the little, short term goals, our most important needs from our less important needs. Assign priorities. Make schedules: which needs should be taken care of first? Which goals do we go after first?

(4) We will get advice where we need it.

(5) We will make sure we have the power to carry out what we decide to do. The decision to marry is a decision made with another person. It is not merely to become married but to *live* married, to *work* and *play* married, to *talk* married, to *pray* married, to *grow* and *develop* married, to *change* married, to *suffer* married, to *accept* and *reject* married, to *fail* and *succeed, win* and *lose* married, and finally to *die* married.

High school students correctly identify irresponsibility, drinking, adultery and temperamental conflicts as four dangerous disorders which can bring a marriage tumbling down. In statements below and elsewhere in the book they vigorously reject the heavy drinker, the playboy lover, and the person with no life aim or ambition as a person suitable for marriage.

The statements below were taken from taped interviews. Imagine you are a marriage counselor listening to these different people telling how they made up their minds to get married. What's right or wrong about each decision?

Statement A: "I fell in love with the pain in his eyes and his unruly hair. He didn't have a job and wasn't sure what he wanted to do. He was so lost and lonely. He needed help and I was the only one who could help him. . . .

STUDENT ANSWERS:

Well of course its totally impractical, she should have *known.* She probably knew he was worthless but still she was in love. She thought with her heart, not her head. But I don't care what any marriage counselor says, I commend her first action. She wasn't one of these people who "don't want to get involved, don't want to get hurt." Well, she got involved and now she's hurt but love always means suffering. Its just a matter of the love outweighing the suffering. In this regard, it didn't.

Girl, 17, St. Louis

Misery isn't love. Irresponsibility isn't love. Loneliness isn't love. More easily help him by teaching him to grow up than by adding new responsibilities to the ones he already has and won't accept.

Boy, 18, Denver

After a while your motherly instincts will wear off and then you will be stuck with a painful eyed, unruly haired, lonely, job hunter.

Boy, 18, New York

He is just saying he needs you because he is too lazy to get out and help himself.

Girl, 19, St. Louis

She sounds like she wanted him as a son not a husband.

Boy, 18, Milwaukee

I think there are other ways to help people besides marrying them. Is this really love? What would happen if he got his hair cut?

Girl, 17, St. Louis

Why commit yourself to a future of pain, hard work and suffering because your romantic, handsome, *failure* needs someone *to take care of him*? Grow up.

Boy, 18, Milwaukee

A dog could fulfill all these characteristics. He doesn't sound like a man ready to lead his children.

Girl, 18, St. Louis

Most girls also find out that it doesn't work even in a dating situation.

Girl, 17, New York

My best friend also used this excuse to go with this boy whom we both knew. It's fine though to have sympathy and want to help someone but not to the point where you get too involved to help him objectively. That's why guidance offices and vocational services are around. Advise him to use these opportunities.

Girl, 17, New York

Well I hope you two have a happy pity party for a life.

Girl, 17, New York

ADVICE
ON ALL
PROBLEMS
OF
LIFE!
LOVE
MARRIAGE
HEALTH
BUSINESS

You should look a little deeper than unruly hair, and find out why all the pain.

Boy, 17, Denver

Helping a person and marrying a person are two different ways of solving a problem. Marrying him in his condition only added to his troubles.

Girl, 18, St. Louis

The girl's motherly instinct took over her thinking.

Girl, 17, Milwaukee

She wanted him to be dependent on her.

Girl, 17, St. Louis

Why were you the only one who could help?

Girl, 18, New York

Statement B: "I met Jill in a Harvest Moon Ball contest. Everybody said we looked so good together we ought to get married. I got along with Jill all right but I guess what really made up my mind was that so many people kept telling me how right we looked together."

STUDENT ANSWERS:

How well you look together?!!!—How about how you *talk* together, you know, like communicate?

Girl, 17, New York

Looks—what the hell are looks?—marriage isn't something which you enter into because you look good together! Marriage is something you enter into when your inner selves look good together.

Boy, 18, Milwaukee

You're kidding me. I hope you look well together at 65.

Girl, 17, St. Louis

He must love her and be willing to die for her and live for her for the rest of his life, this is ridiculous.

Girl, 18, Denver

If someone told him he would look good hanging on a cliff would he do that to?

Girl, 17, St. Louis

Just because a motor looks good in a car this doesn't mean it's going to run.

Boy, 18, Denver

You can be a good listener and still be openminded and considerate in making your own decisions.

Girl, 18, St. Louis

If you're going to marry for looks, before you do look at her grandmother.

Boy, 18, New York

He didn't make up his mind, society did.

Girl, 18, St. Louis

How are they going to look together when they're 50 or 60. When their hair's gray and their looks not so perfect? What's going to happen to their marriage then?

Girl, 17, Milwaukee

I think this is OK. Because if he is a man and kept dating you, he had some feeling for you, and he could stand on his own two feet. He wouldn't have married you if he didn't love you.

Girl, 18, Milwaukee

Marriage cannot survive just on appearances and it may take a lot of paint to keep it looking nice.

Boy, 17, Denver

Plan ahead—how will you look together twenty years from now? You won't always have a Harvest Moon Ball.

Girl, 17, St. Louis

It doesn't make any difference what you look like together, but how you feel together. Love is based a lot on emotion, not physical appearance.

Girl, 17, New York

If the statements can be trusted, almost all of the fourteen hundred and fifty students who answered will one day choose partners wisely and enjoy good, normal marriages. But there's danger in being too smug and cocksure that we're going to decide right. It's easy to make wise decisions now on paper in school. But when we meet someone we love, it's sometimes hard to see them straight.

The Chicago marriages studied by Thomas appeared to be normal at their start. Thomas divided his six hundred cases into two groups, a bad-start group and a good-start group. The bad-start group was made up of "war-bride," "forced-bride," and certain other kinds of marriages which were headed for trouble from the way they started. They made up only 20 percent of the total number of cases.

The bulk of Thomas's cases, 80 percent of them, were marriages which appeared to have started well. Few of the people who threw rice at these wedding couples could have predicted their marriage would end in a court. The brides and grooms all seemed so well matched. They "looked right" for each other.

What Facts do We Have to Deal With?

The first step in the marriage decision is to get the "facts" together. The facts are *legal, economic, social, religious, sexual* and *psychological*. We are more accurate if we call them *realities* instead of facts. Whatever we call them, they have to be faced and dealt with. Whether we make a *mature* marriage decision or not will depend on how we cope with these realities.

The first three realities stand *outside* the decider. These realities are established by the Church, the State and the Community.

REALITY #1. MARRIAGE IS A LEGAL CONTRACT SIGNED IN THE PRESENCE OF WITNESSES BEFORE A PROPER AUTHORITY.

Family stability is important for community progress and health. The state is responsible for guarding community health. It is supposed to keep us from walking into and out of marriage as if marriage were the Book-of-the-Month Club. It asks our age, whether we are citizens of the country, whether we are blood relatives, whether we are being forced into marriage. According to law, wife and child are entitled to the man's support and protection. Once we've signed, the state will hold us to our contract.

THE WEDDING CEREMONY

The wedding ceremony is the *publication day* of a new family.

It is a little *drama* carried off in full costume. Its leading characters speak their lines promising to be faithful to each other for life. The father of the bride acts out his consent to give up his daughter. By his act, the father commits his family not to meddle in the new family's affairs. The drama thus symbolizes both the parents' and society's approval of a new family's independence. The drama ends with a fanfare of car horns and dragged tin cans.

Is the wedding ceremony a frilly, sentimental soap opera, a family spectacular, a day dressed up to be remembered? Is that all—or does a wedding have solid, more serious meanings?

> Religious ceremony is not for show primarily but to impress the couple and all watching with the solemnity, sacredness, spiritualness and importance of a wedding. It gives the couple a feeling of security that their marriage has been witnessed by God.
>
> *Boy, 18, St. Louis*

> Most Catholics want to have Christ in their lives and so desire to be married in Church with Christ as the minister and their guardian in the future.
>
> *Boy, 18, Milwaukee*

REALITY #2. MARRIAGE CREATES A NEW ECONOMIC UNIT.

Who marrieth without money hath good night and sorry days.

English Proverb, 1670

Budget experts say that it now costs the average American man or woman more than $20,000 to raise a child to the age of eighteen. It will cost an added $10,000 to put him through four years of college.

Every new marriage delivers a new buying and spending unit into the economy. It is expected to pay its own way.

If the couple can't pay its own bills, it usually asks for help from in-laws. Here is a source of trouble, for when in-laws help finance a marriage they expect to have something to say about how it is being run. Their decisions may contradict those of the new married couple.

Robert Gibson, speaking for the National Committee for Education in Family Finances, suggests: "Before marriage, a couple should work out a balance sheet covering income from all sources and realistic weekly or monthly expenses. Know in advance what your rent is likely to be and what, for instance, food will cost you in the supermarket. The reserves put aside should be beyond the cost of the honeymoon, regular expenses, and the cost of establishing your new living quarters."

Clark Blackburn, general director of the Family Service Association of America, adds this advice: "After estimating what it would cost to furnish their first apartment or home, a couple contemplating marriage should have enough savings to equal at least three months' income."

I want a good provider with ambition, imagination and courage to face today's tensions.

Girl, 17, Denver

REALITY #3. FOR CATHOLICS, MATRIMONY IS A SACRAMENT.

According to Catholic teaching, Christian marriage is a sacrament by which a Christian man and woman, in union with Jesus Christ and relying upon His grace, pledge to each other a life-long love and fidelity according to God's purpose for marriage.

Catholics believe that this sacrament calls a husband and wife to re-create and act out the spirit of Christ each in his own soul and in the soul of the other.

I'd expect to have a spiritual togetherness, deeper than just going to Mass on Sundays.

Girl, 17, Denver

Religion is necessary to the basic nature of a person and if there is a conflict here, then the two peoples' whole concept of life may disagree, and unless there is a common outlook on life, the marriage will be in danger.

Girl, 18, Milwaukee

The love of each other should not be between only the two, but three persons, you, your mate and God.

Girl, 17, Denver

Technically, a sacrament is a symbol instituted by Christ to give grace. What is the symbol of matrimony? Is it the entire church ceremony, Nuptial Mass included? Is it a certain prayer said by the priest? Is it the priest's blessing?

None of these.

The marriage contract is a two-way pledge of life-long love and fidelity.

Catholics believe that when this pledge is made by a Christian man and woman, it is the *pledge* or *contract which is the sacrament of matrimony.* As a sacrament, the contract is a sign and source of grace.

Grace helps the marriage partners to meet the terms of their contract. What is that contract and its terms? They have promised love, respect and fidelity to each other till death. Grace will help them meet this promise. Grace will help them meet the duties, the obligations, the rewards and joys this promise puts on them. Their state in life now is that of two individuals each bound to anticipate and work for the other's good. Their state in life is that of an independent family. Their state in life requires important decisions. It must grow through dialog. Grace assures them of help.

Dialog and decision-making are the central means for marriage growth. Grace will help overcome all threats against marriage growth.

Dialog requires that individuals meet each other open and unafraid; that they drop their proud false fronts, their coverups, their fake accents, their sneaky maneuvers trying to win petty selfish advantages, that they reveal to each other the undefended workings of their minds. Grace will help them.

It will help as they try to avoid panic and despair or to control anger or destructive haste in making their new family decisions.

It will help as they try to interpret and organize their new goals and as they honestly estimate their abilities to achieve them.

It will help open them to accepting advice.

The marriage contract is pronounced once before the priest but is renewed and strengthened with every little decision and act of fidelity made during marriage. Every act of fidelity carries the generative power of grace. The more one cooperates with grace, the fuller, more vigorous is his life in Christ.

THE LIFETIME MARRIAGE.

> I think a man and a woman should chose each other for life, for the simple reason that a long life with all its accidents is barely enough for a man and a woman to understand each other; and in this case to understand is to love. The man who understands one woman is qualified to understand pretty well everything.
>
> J. B. Yeats

U.S. magazines and newspapers sometimes give the false impression that a majority of our families are stumbling toward a complex breakdown through divorces and separations. The high rate of U.S. divorces is disturbing but no cause for panic, according to Paul H. Jacobson, Ph.D. Dr. Jacobson is author of *American Marriage and Divorce* and is a highly informed person with regard to statistical trends in family matters. His image of the American marriage is optimistic and hopeful. The number of twenty-five-year and fifty-year wedding anniversaries in this country has been steadily rising since 1932. On the basis of this rate of increase, Dr. Jacobson predicts that the number of silver wedding anniversaries will climb from seven hundred and seventy thousand couples in 1960 to about nine hundred thousand couples in 1970 and golden anniversaries from one hundred and thirty thousand in 1960 to about one hundred and eighty thousand in 1970.

"The longer span of life enjoyed by modern Americans," Dr. Jacobson assures us, "has more than compensated for the climb in divorces. For every marriage that ends in divorce, there are three that end in death. The *average life expectancy of a new marriage today is thirty-one and a half years.*"

> The more I study the subject, the more apparent it becomes: marriage is regarded as—and is—the happiest, healthiest, and most desired state of human existence. We get divorced not because we don't like marriage, but to find a better marriage partner. We live longer and are healthier if we are married. More of us

are getting married, more of us are staying married longer and we are getting married younger. Marriage is the central fact of our lives.

Dr. Paul C. Glick of the
Census Bureau quoted in
"This U.S.A. An Unexpected Family Portrait of 194,067,296 Americans Drawn from the Census."

Catholics marry knowing their vows are permanent and that valid Catholic marriage cannot be dissolved.

How does this image of marriage for life affect the marriage decision? It scares certain persons away from marrying at all; some of these might have made good marriages, some would certainly have not. It saves some persons from impulsive marriages which they would later regret.

How does the conviction that marriage is permanent affect persons already married? It can be a source of steadiness and confidence, or it can encourage a person to get careless and over-confident. It can inspire patience and impatience. It can be the reason for peace, security and great satisfaction or it can hang like a heavy lead collar around a person's neck, contributing to feelings of defeat, depression and despair.

The conviction that marriage is permanent may have good or bad effects. Whether it makes a marriage fly or drag will depend on whether the marriage is a healthy, growing marriage or not.

. . . at the end of what is called "the sexual life," the only love which has lasted is the love that has accepted everything, every disappointment, every failure and every betrayal, which has accepted even the sad fact that in the end there is no desire so deep as the simple desire for companionship.

Graham Greene
May We Borrow Your Husband

Almost all students said they wanted marriage for life:

I want always to keep the family together no matter what happens and to this I must be sure to marry the *right* man, as I am a Catholic and I am bound to stay married to this man.

Girl, 17, New York

Faithfulness upon both parts is not important. If you are willing to stick by each other during all kinds of crisis you have marriage licked. The willingness to sacrifice will usually save and protect a marriage.

Boy, 17, Denver

MIXED-IMAGE MARRIAGES MEAN TROUBLE

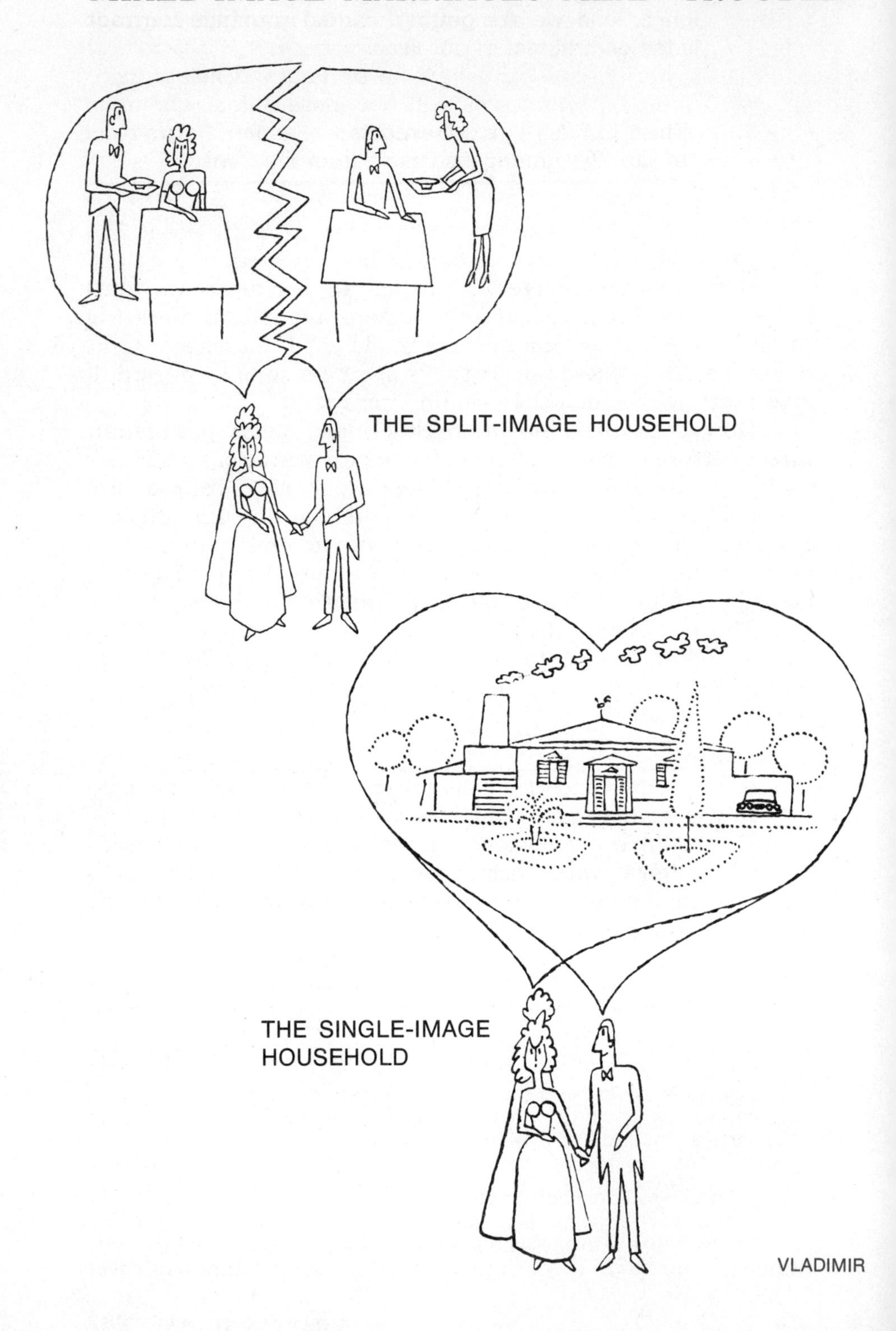

But the thought of a life tied close to the wrong partner worried some boys. Several thought the ideal marriage contract ought to carry some exit hatch for getting out if things got bad. Their statements don't have to be accepted as truthful opinions. Some boys and girls will talk about the liberal marriage they intend to have but when they zero in on the persons they want to marry they become arch-conservatives.

> Marriage vows should contain an "Escape Clause" or the vows should be dispensable. The husband and wife should be fairly faithful, however, should they get the urge I do not think a wee bit of extramarital sex life would hurt. I think this because I believe that probably 25 percent to 50 percent of couples do this anyway. So it would solve many problems to legalize it. In marriage a person should expect kindness, attention, consideration, work and last, sex. The persons must be mature in their outlooks and not be jealous of each other. The marriage vows people uphold now are built from sentiment and tradition. With regards to children after the first few of them (four or five), the parent will be too busy to really worry about other things.
>
> *Boy, 18, Denver*

REALITY #4. IMAGES AS REALITIES.

The realities discussed above exist *apart from* and *outside* the decider. There are *inside, pyschological* realities too.

These are the *images* which influence what the decider will do and whether he will act with pleasure and joy or with annoyance and anger and sorrow. The images represent his Self to himself or to others; they represent the marriage partners and they represent the marriage.

The married person must create a *new self-image*, one in which his Self is intimately related to another person. From the moment of marriage, he will see himself completed by someone else. He will see himself one with her; she will see herself one with him.

Images change from person to person. Some are established by the decider himself. Some were impressed in him by influences which began before he was born, by his early home environment, by brothers, sisters, school friends. They are "inside" facts. They are realities to be reckoned with. Some may have never been questioned before. They will stubbornly resist a reckoning.

How do we know what our images are? We ferret them out with questions.

For example, I want to determine my image of permanence in marriage. I ask myself: is marriage for life or for part of life? Will I stay with my partner under any and all conditions or, back in my mind have I set certain limits to what I'll take? Will I stay with a drunkard? A drug addict? A homosexual? A woman who is cold? Will I stay if his or her in-laws move in and take over?

Will I forgive infidelity? What if there are no children—will I quit then? We promise to stay in sickness and in health. Does that include mental sickness? Emotional sickness? We promise "till death do us part." Does that mean we can quit when communication dies, when the talking stops?

The most important image to analyze before marriage is my image of the marriage contract itself. If I could write my own contract right now, what would I write? What would I vow to give up and what would I expect to get from marriage?

How do I see myself as a husband/wife? As a wife, do I expect to continue college or work? As a husband, do I see my wife going to college, or staying home, or working?

What does love mean to me? Is love *proved* only by action? Is it enough to show my love by working and bringing home a pay check? Have I ever heard my father tell my mother "I love you."? Did I hear it only after they had a fight? Or was it a common thing to hear? Did it embarrass me? Did he mean it?

A study of G.I. marriages in South Vietnam showed that U.S. soldiers who married native women resented and feared American women because they were "demanding," "aggressive," "domineering," "only interested in money and position." On the other hand, these men found South Vietnamese women were "kind," "compassionate," "sincere," and "generous,"—and available.

Capt. William F. Kenny, an Army psychiatrist, and Capt. Albert Kastl, an Army psychologist, conducted the study. They said that most of their G.I.'s who married native girls came from homes broken by death or divorce. Many of the G.I.'s had been divorced. Many suffered sexual inhibitions.

The study surveyed more than one-third of one hundred and eighty marriages between June, 1964, and November, 1966. It showed that a large number of the men had lacked strong or responsible fathers during their childhood. Since they could not *identify with* any men, they identified with their mothers or other female figures.

As a consequence, they now looked for wives who had motherly qualities. They made poor husbands, generally speaking. They expected their wives to be as long-suffering, tolerant, patient, ever-forgiving and ever loving as their mothers had been. It did not occur to them that their wives might have needs of their own which required satisfaction too.

A control group of sixty-four single servicemen disagreed with both views. The Vietnamese women were "cute," "perceptive," "smart, yet crude and uneducated—almost like a three-year-old child, back in the States." The control group liked the American woman for being "full of ambition and able to accept life as it comes—able to stand by me and accept my decisions."

Quiz

1. Describe the seven facts to be considered before marriage.
2. Explain how money affects marital decision-making.
3. How can money affect marital communication?
4. A sacrament is an external sign. What is the sign in the Sacrament of Matrimony?
5. Why is it important to know one's images of marriage before marrying?

REALITY #5. MARRIAGE IS A SEXUAL RELATIONSHIP.

Certain realities are neither wholly outside the decider nor wholly in his mind. They are inter-personal, between husband and wife. The sexual relationship and dialog are two inter-personal realities which are part of marriage.

Sex is a unique force. It drives us to meet each other and to date and to marry. Some persons use sex as a bribe, some use it as a weapon, a means for gaining vengeance. When a sexual relationship is right, it can help a marriage grow. When it is wrong, it can stop a marriage dead.

Sexual intercourse brings a reward of intense physical pleasure. To obtain this reward, some persons will lie, steal and kill. Many persons, when they are told they have to abstain from sexual intercourse or deny their religion, choose to abstain from religion. The need for sexual presence is often the source of human loneliness.

Every individual person has his own unique loneliness. Loneliness is not dead or static. It grows as the individual grows. It grows with the development of qualities which make the person stand out from others. It develops as a person separates himself from the common, as he finds what is special and unique.

Every distinguishing difference and every rejection of some common idea or symbol or behavior separates the person from others. These differences and rejections serve to identify the individual. They mean freedom from the undefined masses and categories.

But freedom and individuality are lonely and lonely man is restless. He paces his island individuality and looks with longing to where the sounds and lights are, for sounds of work and play mean people.

What drives him to leave his "island" is his need for human nearness and love. What helps him leave it is communication.

The center of man's individuality is the Self. Here at this center is the source of dialog. Dialog is man's most perfect means for commuting between his island Self and mainland man.

Affection is a lesser kind of escape communication. Sexual intercourse is the *grand act* of all human touch communication. It is the most intimate form of body to body nearness.

But sexual intercourse separated from dialogic love is a crude and unsatisfying form of communication. Unless there is dialog, when the act is complete and body separates from body, the person is as lonely and isolated as before. Without dialog, intercourse is merely a physical act more or less exciting and pleasurable. It brings intense physical pleasure and relief, but without dialog it brings sadness and dissatisfaction. It cannot relieve man's deeper loneliness or take him from his island.

REALITY #6. MARRIAGE IS A DIALOGIC RELATIONSHIP.

Marriage makes a new kind of being. Two separate independent persons agree to become biologically, psychologically and economically dependent on each other. They share new composite goals and new possibilities of growth. Decisions must be marked not only "his and hers" but "theirs" too. The decider must consider another's future and another's needs too. He now must make family decisions. Decisions are often made by two deciders acting jointly. All members of the family depend on the decider's judgment and his or her ability to communicate it.

Marriage changes the *I* in *I and Thou*. The *I* is no longer alone and independent. It is *I and another and Thou now*. I and Thou by a kind of mitosis becomes *I-Family-Thou*. This is true even when the *I* is husband and the *Thou* is wife. In the husband-wife dialog, both partners are committed to their marriage contract. Neither husband or wife can speak from their former single-state condition. When either addresses a person outside the marriage, he speaks as a married person. When both speak to each other they speak as married persons too.

The I-Thou communication of single life is simpler than the I-Family-Thou of married life. The single person can break off communication when it suits him to break it off. He can stage his dialogs, carefully choosing his sites, the time and the persons he will talk with. He can pick subjects to discuss that please him while avoiding undesirable subjects. He can quit when he is tired of them or they of him.

But in marriage, the two persons are tied together. There is no picking the best time. There is no waiting till you feel right. There is no walking out, no getting away.

Non-marrieds can get a rest from dialog. There is no rest in marriage. When one dialog is put down the next one is being readied. Trust and belief are the foundations for dialog. These foundations need constant attention for they stand on shifting grounds. Persons who live together have hundreds of opportunities to strengthen or weaken their confidence and trust in each other.

Language is the only medium for most dialog and silence is usually a retreat or rest. But as married persons come to know each other, symbols, gestures, facial expressions also become part of their dialog. Their is no retreat to silence because as silence itself becomes noisy with meanings it joins the vocabulary of dialog.

Outside of marriage, partners in a dialog usually have their own separate goals and needs. Frequently dialog consists in the gradual revealing and discovery of these goals and needs. The temptation to falsify them is strong.

Take a young man and woman who meet. She's twenty-one and wants to marry as soon as she can. He's twenty-three and has career plans to settle before he even intends to think about marriage. She likes him, but is afraid if she says she wants to get married he'll bolt so she lies and says she's not interested in marriage. Or she tells the truth and he lies by letting her think he's seriously interested in her.

Marriage partners share the same goals and needs. Certain kinds of information kept back by one of the persons may be an act against dialog. Suppose a man is frightened to tell his wife he can't make it on his job. Or suppose he's fired and spends the rest of the day in a bar instead of going home with the bad news. Suppose a wife is afraid she has cervical cancer and is afraid to tell her husband and afraid to tell even her doctor about it. (Every year, fourteen thousand women die of cervical cancer. According to Dr. J. Ernest Ayre, the cancer cytologist, many of these deaths might have been prevented had the women submitted to early examination.)

In pre-marriage relationships, sexual differences often distort and distract dialog. They are the source of fakery and masks. They create feelings which rattle the attention. Mere physical pleasure is often the masked objective of boy-girl meetings. This objective is kept secret because the persons fear rejection or frustration should their true yearnings become known. But in marriage, these sexual tensions are satisfied by intercourse. The fuller needs of the person can now be sought and answered. Intercourse or withdrawal from it frequently starts dialog.

Sexual intercourse in marriage may serve dialog by breaking down the pride systems which interfere with dialog. It may help reduce tensions related to a poor self-image, particularly a poor body-image. Since it is an act of intimacy and privateness, of special, privileged nearness, it is the material of trust. It is an act which explores and exploits differences. It is an act of mutual acceptance, the far opposite of rejection. It assuages loneliness without destroying individuality.

Even enemies can dialog. But in an environment of hate, suspicion, distrust and misbelief, the progress of dialog is arduous and its success is meager. In a mature marital dialog there is an environment of love, trust and belief, of willingness to sacrifice the self, of willingness to serve the good of the other. Dialog under such conditions can't help but be richer.

In marriage less and less effort is needed to unmask and de-falsify the *I*. An authentic *I-Thou* relationship becomes more and more a matter of habit.

All persons who dialog are rewarded by growth and *becoming*. When single persons separate and go their own ways growth usually sputters out. The rewards of growth from marriage dialog don't stop. Rather, they are compounded and increased. Partners continue to explore and draw out newly discovered sources of *becoming* in the Self.

At the same time, intimate, day by day, and night by night living puts intense pressures on communication; sexual intercourse, pills, babies and in-laws affect communication; the falling away of old friends affects it; liquor and "other women" or "other men" affect it too. But through all changes dialog must persist.

If this seems hard, the rewards of persistence are great, for where dialog endures there is growth, there is *becoming*, there is *being*, and there is happiness. Don't run from dialog. Give up. Surrender to it as to the Hound of Heaven. You can't lose.

REALITY #7. MARRIAGE CREATES THE FAMILY, A NEW SOCIAL UNIT.

Marriage gives man new goals offering greater possibilities of *becoming*.

Individual man *becomes* through art and work. He *extends his being* through business, through the sales he makes, the orders he gives and executes. He may produce books or plays. He may teach, serve tables, manufacture shoe soles.

But the average married man can trust his being far beyond the average single man's limits. Married man's Potential Self grows through persistent dialog with his wife. Beyond their mutual *becoming*, both husband and wife *become* more in their children. Beyond their children, their image is thrust into the community that will one day be served by their children and beyond that community to the nation and beyond that to mankind and history.

Married partners record their decisions in their children. They impress their children with their attitudes towards others, their tastes, their habits of judging and communicating, their ways of solving problems.

Civilization is a stream so broad, so deep and so long that it seems sometimes we can have little or no effect on it. We feel it will go on with or without us. Just as our single vote won't likely change an election, neither will our 3.6 children change the world for better or for worse.

This is an unimaginative forecast, a grim, niggardly cynicism. It contradicts the spirit of Jesus who looked down the hallways of centuries, who made decisions which changed mankind, who spoke a timeless dialog to all of us in any age.

The purpose of marriage is not merely companionship and mutual Self-fulfillment. Nor is it merely to have babies. Nor is it enough to have babies and educate them if "educate" means merely to put them through Catholic schools. There is a purpose to education and it is this purpose which must guide our child's education.

If we consider that our habits and attitudes, our defenses and aggressions were learned for all our lifetime in the "university of the home," then it becomes clear what we as parents ought to teach our children. By our example and by direct instruction we ought to show them how to make proper decisions and how to dialog with the community of man.

Our children will learn to use the instruments of *becoming* by watching us. They will see how we relate, how we love, how we meet their meanings, how we listen, whether we dialog with each other and with them. They will observe and learn from us how we judge the world. Formed this way, they will go out to meet that world.

Economists, military experts and welfare agencies are concerned about our population's size. They count its numbers and keep estimating how much it's going to eat and spend. Governments worry about *quantities* of people. It is up to parents to worry about population *qualities*.

Some parents pollute the population stream, others clarify and enrich it. But childless man doesn't contribute to it at all.

Know Your Feelings

DON'T DECIDE WHEN YOU'RE HIGH.
DON'T DECIDE WHEN YOU'RE LOW.

We know true love lasts forever, ideal love fulfills, heroic love is unselfish, etc. We want our love to be true, ideal and heroic. But how do we know when it is?

But what is it when something is *not* true or ideal or heroic? We don't know for sure. We say we wasted time. We say we were "infatuated."

Fatuous (pronounced fat-chew-us) means silly, stupidly unrealistic and empty.

The word *fad* may come from the word *fatuous*. A fad is a custom, a saying or expression, a gimmick or game or person popular for awhile and followed with exaggerated, intense excitement, then suddenly dropped. You're sick of it. Only the square, retarded and out still hang on.

"Ring-a-ding-ding," hoola-hoop, the twist, Chubby Checkers, the skinny necktie, psychedelics, stretch pants, dyed fur, granny dresses, the miniskirt, bell-bottom slacks, white Correge boots, Sonny and Cher, Batman, the frug, the swim, the monkey, Scrabble, white makeup and black eyes, Herman's Hermits, Lolita sunglasses, mink sneaker laces, The Animals, yo-yos, Sal Mineo, knee decals, black nailpolish, beehive hair, squirrel tails on bikes, groovey, the Hippie movement.

These are fads.

If you're "in love" with a new boy or girl this month, last month's love was a fad—a person followed with intense and exaggerated excitement till suddenly: DEAD! You can't bear to look at her, or you hate yourself for letting him touch you. How could you do it.? Yettchhh!

Calf love, Half love.
Old love, Cold love.

Old English Rhyme

It's easy enough to look back after being disappointed in love and say it was a mistake or only "sexual tension" or a temporary craving for affection. But no one can see this from the middle of the feeling. To the person caught up in an "infatuation" these feelings are very real. Calling them silly won't make them go away. That's hindsight wisdom.

Persons under twenty are still going through important biological and psychological shifts. As their bodies change and their personalities adjust to adult life, their needs shift and change too. They shift from one sense of isolation and loneliness to another and then still another. They're not usually ready for long-term love because they don't exactly know who they are. And if they don't know who they are, they know still less who their lovers are.

The one thing we can tell them is: hang on. Wait it out. If a love is worth anything, it won't likely go away. If it does go away or fails to grow you're ahead. At least you haven't committed yourself.

We may not always be able to describe the precise early warning signals of an infatuation but we can describe mature love. We can describe an emptiness only by what is supposed to fill it.

To love is *to think, to will, to do* the *good* of *another.*
Cana: Denver

Love follows knowledge, self knowledge and knowledge of the other; knowledge of both actual and potential self. We never fall *victims* of love unless we let ourselves. We love because we decide to love and in the manner and degree we decide.

The more you know about a person the more you can love him in spite of his defects. I believe that no one should marry unless he can point out three major shortcomings—as well as virtues—of his intended partner. You can't learn that during a short engagement.

Msgr. William F. McManus
Family Life Bureau of the Catholic Archdiocese of New York

Sexual attraction is the longing for body to body nearness. Love is a longing for full Self to Self nearness.

If you love an animal, you pat him, stroke him and feed him. Human beings need more.

Your communication cannot be limited to touch. You will strive to define your whole Self so that the one you love can know you and share your own *becoming.*

Love doesn't fake. It doesn't let *seeming* creep between the Self and the Other. Good or bad, the person wants to be known as he truly is.

Love respects the full person and meets the person fully and openly in dialog. You don't use a person you love as an object, a thing, an instrument to please yourself.

You love not because of the tingles you feel but because the person you love is worthwhile and deserves to be loved.

TEN COMMON ERRORS ABOUT LOVE.

1. That to love is easy; it comes naturally, spontaneously.
2. That love is something that happens to us; hence beyond our control (e.g., "to *fall* in love").
3. That love is essentially an emotion, a feeling.
4. That to love a person you must also like him.
5. That self-love and selfishness are one and the same.
6. That Christian love involves only a denial of self, a diminishing of self (not a fulfillment of self).
7. That to be loved is always more important than to love.
8. That love may be primary in a woman's life, but should not be so in a man's.
9. That love of others and success in a competitive society are incompatible.
10. That should one "fall out of love," he no longer can love this person.

Cana: Denver

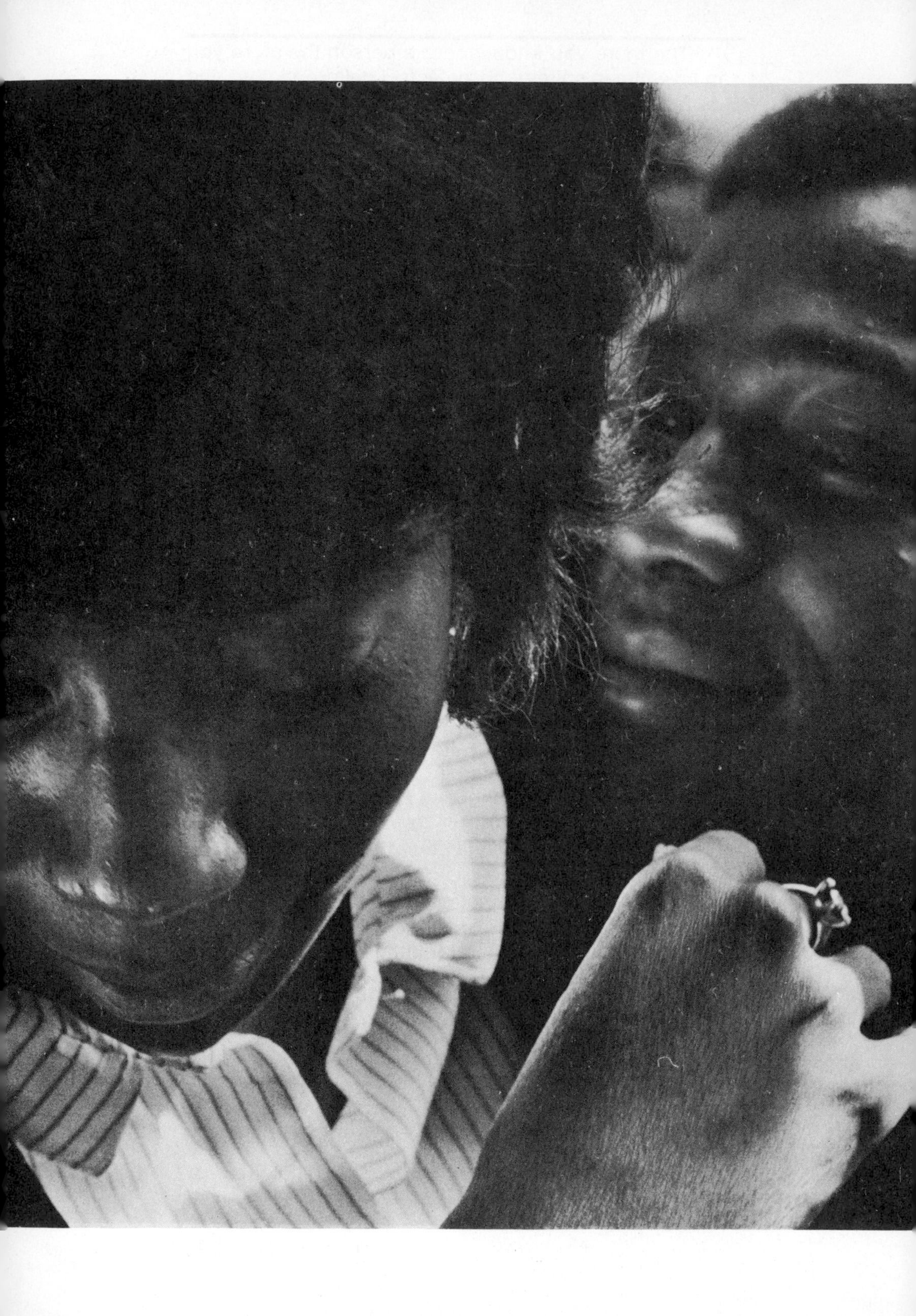

THREE *CRISES* OF LOVE:

1. Novelty fades (sense excitement dies down, thrill of discovery dies out).
2. Disillusionment sets in. (In the early stages, the lover tends to project into, to see in the one loved his or her ideal; the one loved tries to appear always at his or her best.)
3. Sacrifice is called for. Selfishness—the continuing problem comes to the surface. (Real love does not make us selfish; It consistently reveals to us how selfish we are; how insecure, possessive, jealous.)

Cana: Denver

Love grows through the normal channels of growth, that is, through judgments or decisions and through communication and dialog.

To love someone is to want his *becoming*. You want yourself to be the agent of that *becoming* in the other, but you will recognize there are other sources. Love is not jealousy. You will want your own *becoming* through the person you love.

Love is *open*. It believes and trusts. It doesn't judge too quickly. It doesn't keep looking for signs and proofs. Love doesn't try to win or beat the other. It isn't competitive.

There is no greater invitation to love than loving first.

St. Augustine

In your thinking, your judgments, the smallest and most secret, you will decide *for* the other person's good, for the other person's *becoming*. Love is authentic. It strives outward from the origins of the Self and reaches towards the origins of the other's Self.

Love is generous. Love thinks of giving. Love sings. It can't stand being hidden. It wants to tell about itself, to spread, to help things grow or get better. It's hard to keep love a secret.

Four things cannot be kept close: love, the cough, fire and sorrow.

James Sandford, 1572

Sometimes a person's Actual Self doesn't seem lovable. It may bore or offend us. In this instance, we love by ignoring our feelings, not letting them force our judgment. We put our hope and belief in the person's Potential Self. No missioner would bury his life in a jungle or ghetto if he didn't

believe in the redemption of the Potential Self. The Alcoholics Anonymous and The Salvation Army would never have been formed if their founders had not believed and loved the potentially lovable human being who is miserably buried in the alcoholic.

The greater a man or woman's pride, the less he or she can love. The profoundly proud person is incapable of love. Why is this so? Because, absorbed in self adoration, he is unable to enter into an authentic personal union. He cannot be really aware that his spouse truly exists as a distinct person, with a value and needs of her own. He cannot, admit to a need of this other as a person—since he denies that he himself is a limited person. And finally, he cannot escape from this prison of self and relate to this other, person to person, on anything that approaches equal terms. To repeat: people who are deeply proud cannot sincerely love. But if husband and wife are to become full persons, and communicate—love is a must.

Francis F. Bakewell, S.J.
Dialog at Cana

What is Your Image of Love?

I expect to share myself with him but not be disappointed if he doesn't understand all of it. I will accept him as a unique person who loves me but is not perfect and I'll expect the same acceptance from him. The most important thing I'll give is my love emphasizing understanding and acceptance not tolerance. I want him to love me. And together I want us both by giving to make our love perfect . . . love requires suffering but we should still be able to say we truly love each other.

Girl, 18, Denver

Right now I go to fancy places and wear expensive clothes and get my own way pretty much. But if I had someone who really loved me, rich or poor a street cleaner or a President of a company I would try my very best to do all he wants and expects of me and all I would want from him is love and a good place to raise our children.

Girl, 17, Milwaukee

I would give him my love, respect him, comfort him, grow with him, laugh with him, to bring up our children to be like him in his good ways. I should realize his faults and realize he's a human being with a lot of responsibility and give him all the happiness I could. I should and would lead him to a closer union with God if this wasn't done I would give him up. For me I lose or give up nothing, I receive the satisfaction of being loved. If I am not loved, I receive the satisfaction of loving him.

Girl, 17, New York

Children need constant care, but so does a husband. A heart, though, cannot be divided, she merely loves all with her whole heart.

Girl, 18, Milwaukee

To start off, I better explain that most people think of me as an optimist in regards to what I expect in a marriage. One of my biggest complaints about marriage is the arguments over nothing—since I believe that a little spat once in awhile adds to a marriage but constant bickering over lack of understanding really gets me. I think people have to learn to love in marriage, it's more than just climbing in bed at night. It's something that has to grow. If I get married, I want my husband to be understanding and I want him to get the same from me. I feel that if people really love they will always think twice in a situation and try not to hurt their lover. Of course, there may be times when pressure builds up and things can be said that weren't realized in advance, then if the people are really in love with each other and realize what love is and how it works, then they will mend their argument. They tell me making up is very enjoyable. I really think love is very complex and I won't pretend to give any easy definitions of it, because it is something you learn on your own and then you become real through it.

Girl, 17, New York

The kind of married life that I want is one that revolves around love. Not the emotional kind, but true love. The love that is a giving, not asking for anything in return. One that consists of helping each other grow. It consists of a trust and understanding. And actions not just words. A good marriage needs love or else there can be no marriage, marriage revolves around love.

Girl, 17, St. Louis

If there is true love, there will be no need for vows of fidelity, or promises to remain in all kinds of hardships. This love will have a real understanding of the faults of the other party and the ability not to let them interfere in happiness. Actually, I don't really believe fully in the statement that "love conquers all," but in my opinion it is a big help.

Boy, 18, Milwaukee

I would give my whole heart and mind to my wife along with all the love I could muster up out of this little body of mine. I would give my life to save my wife if the occasion ever arose. God help me if I don't make the right choice.

Boy, 18, New York

A smile can be the most important sign of love, if you truly mean it. It can mean more than being romantic and having passion love all of the time. . . . In return I want a smile when I'm tired, food when I'm hungry, and love lasting until I die.

Boy, 18, New York

I just want love. That's all I have to give and all I hope to get, what else is there? Now, I couldn't give her my freedom but someday I will. I don't want to be changed or change. I want to be myself and let her be herself. We'll be one and head toward that common goal reading existentialist poetry and drinking sweet wine and changing diapers.

Boy, 17, Denver

Ask Advice

KNOW WHERE TO GO FOR IT.

A girl, will spend hours dreaming of the kind of married life she wants to have. The man she'll marry will be a tall, blonde, cigarette model of a man, a successful doctor or lawyer, a leader, a good father and provider, a religious, sensitive man. With this ideal in mind, she rejects suitor after suitor. Then, to everyone's surprise including possibly her own, she marries a short, pink-haired, kettle-bellied person who wears food-spotted ties and rumpled suits, who smells of cigars, garlic and whiskey.

Is he a great intellectual? No. Is he an artist, maybe? He hates art. Then he must be good and kind. Not that either. In fact, he's cruel to her. He embarrasses her in public and beats her at home. She leaves him after a mere hundred days of misery.

Why did she ever marry him? How could she marry someone so different from her ideal? Because of *feeling. She FELT a peculiar need for this particular person. She couldn't bear the feeling of being without him.*

In spite of pressure from family, and friends, contrary to knowledge and personal experience, contrary to all good sense, thousands and thousands of persons each year marry wrongly. They do so because of *feelings buried in their personalities during their first few years of life.* A person's choice of a marriage partner may reveal whether he felt secure or insecure as an infant and child.

Certain psychological difficulties rise from the very roots of our personalities. They are so subtle they may stay hidden even after a long period of careful, questioning courtship. It's good sense, therefore, to discuss marriage with a qualified advisor. A capable marriage counselor may be able to help us detect these difficulties. He may show us how to accept them and turn them to our advantage rather than letting them trip us up.

aguar

Don't Decide Beyond Your Power to Act

When is a person mature enough for marriage? When he or she is ready to meet the requirements of permanent married life.

A mature person usually makes mature decisions. He makes himself reasonably sure he has all the pertinent facts available to him. He doesn't decide impulsively, when he feels high, or in desperation when he feels depressed. He is able to distinguish serious from frivolous objectives. He is open to advice. He makes decisions he can handle. He is able to walk away from a decision once made and not let it nag him haggard. He decides and acts. The immature person decides and dreams of acting.

Dialog is itself a sign of maturity, for dialog is the perfection of human communication.

Married life requires a double maturity, not simply one person's maturity but the maturity of two persons living close together. Their image of love and sex in marriage is open and realistic. They evaluate themselves rather fairly, not too low, not too high.

And I'd like him to be
more interested in me
than he is in himself
and more interested in us
than he is in me.

Marriage decisions are co-decisions. They are made not only for the decider's benefit but for the decider *and another*. They are often made not by the decider acting alone but by two deciders acting jointly.

The facts a married decider must gather before acting pertain to a family, and the goals the decider is advancing toward are to a family's goals. The married decider trying to act within the limits of his or her power finds that family needs have become limiting factors.

One of the first tests of maturity for marriage is the test of money management. Can the couple agree on a budget? Can they work out a common sense priority list for spending money? If he is earning $95 a week and she $65, should they buy a Cadillac or a Volkswagen?

Badly managed money leads to another test of maturity: how well can they get along with in-laws who may be paying their rent, baby-sitting for them, and making the payments on that Cadillac?

In general, persons are mature enough for marriage when they can cooperatively make decisions leading to the health and growth of their marriage. They are mature when they can face up to the demands of marital dialog.

Images: the Mature Marriage

A wife has to be mature and that means no moodiness, self-centeredness. That would mean talking to your husband when you don't feel like it—being interested in him even when you're upset but most of all being open to each other—completely open.

Girl, 17, Milwaukee

Both persons have to be able to compromise, see things maybe a little differently than normally, respect the other persons rights and the other person. And if there is love between the two people not only will you give these things but automatically you'll get them back again.

Girl, 18, New York

I want a solid married life. After going to school for 20 years and receiving a degree let's say in law, I want to be able to have a strong and devoted family to come home to. With all the divorce in the U.S. this ideal of the family I know will be hard to obtain. . . . I want a large family. I think it aids in strengthening the family, and making it more the center of life. All my ideas though are based on a good financial income, and this is what I'm struggling for now.

Boy, 18, St. Louis

I would detest being treated as a fragile doll who might break if not coveted with affection—I want someone who will listen when something is on my mind and with whom I can reveal my inner self entirely but this should be established before marriage—of course a wife must learn to live according to what her husband can give her she cannot expect to have all she had at home—I want my husband to be an individual with his own ideas and his own set of values which he goes by no matter what others may say of him. I expect of him a home and nice things, clothing, food, support and I expect to do my best to keep these things in good order . . . I hope he can respect me for wanting to be a Ph.D.

Girl, 18, New York

In my future marriage I expect to give the all of myself in a living union with another human being, who has chosen me because he wishes to give the all of himself to the one person with whom he has decided he can best grow and continue forward in the remaining years of his life. I want not so much to *give* completely as to share wholly everything God had given me that makes me me, and I want to share with my husband the all of his being. I will give up nothing that has not completely lost all value in relation to my life from the exchange of vows on. I will give all that God has to offer of a life of giving and sacrifice. I expect to grow together with my husband in the many small things that make up one large whole—a life directed towards *our* God. I will not lose my individuality or my principles, but rather become more of a person by living up to the reflection of myself that I see in my husband's eyes.

Girl, 17, New York

I would try to understand him the best I could and would try to put up with some of the things that would bother me. I wouldn't try to change him to suit my tastes. I would accept him as he is. But if he had too many faults or ways that I didn't like I would not marry him in the first place.

Girl, 18, Denver

The marriage contract must be entered into by a man and woman. Not a boy and girl, but man and woman.

Girl, 18, Milwaukee

I must keep the house and cook. Also I feel it is important that I take good care of myself and don't go to "pot" like many women do. I feel I should also get a good education so that I could take over making the money if it were necessary.

Girl, 18, New York

NO PARENTS are to live with or in the same house or a certain area close to it. All things which could hinder the relationship are to be dropped. Each should take an active part in the home. Each should try to participate in each others lifes such as sports, bridge, etc. Each must give wholly of himself to his mate. Everything must be done in full knowledge of each other. No matter what expense, children must be given the best possible Catholic education. Attend church together.

Boy, 18, St. Louis

Quiz

1. Describe seven ways in which a dialog between married persons may be more difficult than that between single persons.
2. Explain: "Marriage offers man greater possibilities of *becoming*."
3. Define love.
4. What is the difference between love and infatuation? Give six differences.
5. Describe eight common misconceptions of love.

Credits

Grateful acknowledgment is made to the following for use of copyrighted material:

Purdue University, Division of Educational Reference, for excerpts from its Purdue Opinion Panel polls.
Doubleday and Company, Inc., *Marriage East and West* by David and Vera Mace, 1949, pp. 144-148.
Doubleday and Company, Inc., (Anchor Book), *The Human Condition* by Hannah Arendt, 1959, p. 213.
American Psychiatric Association, for definitions of terms from *A Psychiatric Glossary*, 1964, pp. 13, 23, 27, 38, 66, 73, 79.
Prentice-Hall, *Psychoanalysis and Psychotherapy* by Robert A. Harper, 1959, p. 170.
Prentice-Hall, *Building a Successful Marriage* by Judson R. Landis and Mary G. Landis, 1948.
Prentice-Hall, *American Catholic Family* by John L. Thomas, S.J., 1956, pp. 217, 220, 224, 225.
Princeton University Press, *Society and the Adolescent Self-Image* by Dr. Morris Rosenberg, 1965, pp. 143, 168, 171.
Psychiatry, The William Alanson White Memorial Lectures, Fourth Series, "What Can Philosophical Anthropology Contribute to Psychology?" by Martin Buber, vol. 20, No. 2, May 1957, copyright 1957 by The William Alanson White Psychiatric Foundation, pp. 107, 109, 111, passim.
Pendle Hill Pamphlet 106, *The Way of Man* by Martin Buber, pp. 13, 14, 32.
Alfred A. Knopf, *A New Dictionary of Quotations* by H. J. Mencken, 1952, pp. 18, 225, 1312, 1318, 1321.
Seventeen Magazine, July, 1965, "Love and Sex" by Dr. Daniel Sugerman and Rollie Hochstein.
The New York Times Magazine, 1966, "Every 6th Teenage Girl in Connecticut," by Ruth and Edward Brecher. Used by permission of McIntosh and Otis, Inc.
The New York Times Magazine, May 28, 1967, "Report from Teeny Boppersville," by J. Kirk Sale and Ben Apfelbaum.
Farrar, Straus and Company, Inc., *Creative Fidelity* by Gabriel Marcel, 1963.
New American Library, *The Boston Strangler* by Gerold Frank, 1966, pp. 313-314.
Rinehart and Company, *American Marriage and Divorce* by Paul H. Jacobson, 1949, p. 170.
Sheed and Ward, *The Illusion of Eve* by Sidney Cornelia Callahan, 1965, pp. 213, 214.
William Benton, Inc., *The Great Ideas Today* by Lucius F. Cervantes, S.J., p.8.
Encyclopedia Britannica, Inc., *The Great Ideas Today 1966*, "The Difference of Woman and the Difference It Makes," 1966, p. 8.
Saunders and Company, *The Sexual Behavior In The Human Male* by Alfred C. Kinsey, Wardell B. Pomeroy and Clyde E. Martin, 1948.

PHOTO CREDIT:
Photos by George Riemer appearing on pages 1, 20, 38, 50, 56, 90, 104, 162, 186, 190, 266, 272, 278, 296, 300. Photo of Martin Buber, p. 118, courtesy of Dena. Photo on p. 294, courtesy of Lawrence Fink. Photos on pp. 236, 276, courtesy of Ken Heyman. Cover photo by Robert Bull.

Index

ADOLESCENCE: 45-46, 99, 171-172
- daydreaming, 170-172, 176
- escape from home, 176
- nubility, 169-171
- puberty, 169-171, 188
- second birth, 45-46
- steady dating, 263
- student opinions of teeny marriage, 254-255
- teenage marriage, 252-253

ADULTERY: 216-234
- alcohol, 220-222
- divorce, 119-200
- early marriage, 217-219
- five "causes" of, 219-223
- and the law, 224-229
- work, 219-220

ADVISORS, TEST OF GOOD, 76-78

ALCOHOL:
- and adultery, 220-222
- alcoholism, 215
- and communication, 209-210
- decision-making, 208-209
- and divorce, 199-204
- effect on brain, 206-207
- and marriage, 203-205
- stimulant/depressant, 205-208
- vs. becoming, 208-209

ASSOCIATIONS AND ORGANIZATIONS:
American Cancer Society, 177
American Humane Society, 202
American Institute of Family Relations, 263
American Medical Association, 215
Arthritis and Rheumatism Foundation, 177
Cana Conferences, 181-182
- at Denver, 293, 296
Census Bureau, 280
Family Court Milwaukee (Proceedings), 74
Motion Picture Producers and Distributors of America, 227
National Committee for Education in Family Finances, 277
National Safety Council, 71-72
Public Health Service, 177
Rutgers Center of Alcohol Studies, 211
Salvation Army, 151

AUTHORS QUOTED:
Allport, Gordon—215; Arendt, Hannah—231;
Bakewell, Francis—297; Baldwin, Alfred L.—82; Blackburn, Clark—277;
Callahan, Sidney Cornelia—184; Caprio, Frank—222; Cervantes, Lucius F.—184-185; 262;
Davidman, Howard—187;
Frank, Gerold—242-243; Freud, Sigmund—33; Fromm, Erich—47;
Gibson, Robert—277; Glick, Paul C.—281; Guerber, H.A.—188-189;
Hoffman, Anna Rosenberg—164; Horney, Karin—47; Howe, Reuel—120;
Jacobson, Paul H.—256-7, 279; Joyce, Alfred—268;
Kinsey, Alfred—209, 218;
Luckey, Eleanor Braun—191;
Mace, David & Vera—28-32;
Neuberger, Maurine—183;
Popenoe, Paul—263;
Rutledge, Aaron—110, 232, 237-238, 241, 243-244; Rosenberg, Morris—55, 114;
Sullivan, Harry Stack—47;
Wilson, Meridith—301.

AYRE, ERNEST J., 287

BSI (BAD SELF IMAGE): 54-58, 59, 61-62, 94, 156, 249
- beating BSI, 62-63
- freezes communication, 54-55
- origins of, 59-62

BSGF (BASIC SENSE OF GOOD FEELING): 44-45
- and touch communication, 143-144

BECOMING:
- alcohol, 208-209
- a bad date is not becoming, 52
- birth, 44-45
- "central drive", 43-64
- in dating, 39-43, 43-64, 183
- a good date is becoming, 51
- enemies of, 197-251
- helped by impressions & expressions, 48-49
- in marriage, 287
- personal, 65-84
- self, 48-49
- social, 65-84

BUBER, MARTIN, 86, 115-121, 126-127, 140-143

CANNY, FRANK, 230
CHANGE, 123-125, 167

COMMUNICATION:
- alcohol, 209-210
- bad news, 132-134
- barriers to dating c., 88-107, 127, 139, 170
- and car, lack of, 101-105
- characteristics of, 86
- common interests, 92-93
- compared with dialog, 126-130
- conceit, 96
- conflict of temperaments, 237
- constructive, 86
- in dating, 1-7
- and double dating, 7-8
- effects of prejudice, 125
- failure, 88-89
- family interference, 97-98
- gestures, 85, 108, 143, 286
- how it starts in infancy, 88, 108-115

—immaturity, 239
—language, 109-115
—the listener, 87-88
—in marriage, 191, 230
—and maturity, 269
—misunderstandings, 91
—money, 250
—money, lack of, 105-106
—moodiness, 89-91
—no place to go, 97
—poor sense of humor, 93-94
—sex instruction, 156-158
—social becoming, 85-158
—steady dating, 264-265
—symbols, 85, 103-105
—touch, 143-145
CONFLICT OF TEMPERAMENTS:
—and communication, 237-238
—and divorce, 199-200
CONIGLIARO, VINCENT, 230

DATES:
—calling off, 131-132
—communication failure on, 3-7
—sex centric, 145-152
—successful, 19-22
DATING:
—analysis, 8-22
—and becoming, 39-43, 43-64, 183
—and communication, 1-3
—double, 7-8
—girls handicapped, 9-10
—history, 28-37
—parents, influence on, 28-30
—roles, 22-25
—single teenager, 25-27
DECISION-MAKING:
—and alcohol, 208-209
—and becoming, 65-84, 182
—guidelines for, 82
—and identification, 80-81
—and immaturity, 239
—and judgment, 66-67
—in marriage, 191, 248-249
—and money, 105-106, 249-250
—rules after, 75-76
—rules before, 71-74
—self image, 81-83
—in steady dating, 264-265
DECISIONS:
—early development of, 78-80
—importance of little, 68-70, 131
—John and Edith McFaum, 68-70
—to marry, 252-303
DEFINITIONS:
—anxiety, 40
—BSGF (Basic Sense of Good Feeling), 44-45
—BSI (Bad Self Image), 53
—body image, 53
—communication, 85-86
—dating, 1
—decision, 67
—defense mechanism, 80
—depressant, 206
—dynamics, 99
—dynamic psychiatry, 101
—ego, 81
—egophile, 55
—egophobe, 55
—fidelity, 229
—identification, 80
—infatuation, 291
—infidelity-chronic/occasional, 218
—judgment, 66-67
—maturity, 267
—prejudice, 124-125
—regression, 81-82
—repression, 99
—self:
actual, potential, real, true, 46-48
—self esteem—high/low, 55
—self image, 53
—superego, 81
—withdrawal, 42
—zenophobia, 163
DIALOG: 86, 119
—anti dialog, 152-153
—barriers to, 130-131
—Buber, Martin, 116-120
—compared with communication, 126-130
—crises of, 132-134
—dialogic person, 141-143
—and difference, 159-195
—dodges, 134-139
—first principle of, 121-130
—and images, 124
—I-Thou, 119-120, 121-123, 130, 152-153
—man & woman difference in, 160-189
—in marriage, 191-195, 230, 278
—and prejudice, 124-125
—rules of, 140-141
—with enemies, 287
DIFFERENCES (SEE ALSO MAN-WOMAN), 123-125, 160-169
DIVORCE:
—adultery, 216
—"causes" of, 197-200
—conflict of temperaments, 237-238
—drinking, 200-215
—in-laws, 238-240
—irresponsibility, 234-235
—mental illness, 245-246
—money problems, 248-250
—religious differences, 247-248
—sexual incompatibility, 240-245
DORFMAN, RALPH, 170
DRINKING: 200-215
—and marriage (student opinions), 212-215
—and maturity, 268
—and students, 210-212

EGOPHOBE, 55-58

FATHER, 32, 82, 151, 181
—unmarried, 151-152
FIDELITY: 218, 221
—creative, 232
—and dialog, 232-233
—in marriage, 279
—reason for, 229-234
—in steady dating, 265
FORD, EILEEN, 231

HELD, JOHN JR., 34
HOMOSEXUALITY, 180, 187-188, 222
HORMONES, 169, 175

IDENTIFICATION, 80-81, 83, 187-188
IMAGES:
—bias, 124-125, 141
—of dating, 22-25
—of decision-making, 81-83
—of fidelity, 284
—of listeners, 113
—of love, 283
—of marriage, 281-283
—of mature marriage, 302-303
—of self, 53-63, 102, 141, 148, 166, 209-210
—of sexual identity (sources), 184-189
INFANCY, 43, 59, 78-80, 82, 88, 109-110, 143, 223
INFIDELITY:
—causes, 222-223
—punishment of, 222-229
—results, 223-224
IN-LAWS: 238-240
—and divorce, 199-200
—and marriage decision, 277, 302
IRRESPONSIBILITY, 234-235
—and divorce, 199-200
JUDGMENT: 66-67
—and sin, 66
—and taste, 66

KENNEDY, JOHN F., 73-74

LISTENING: 86, 87-88, 92, 93, 113, 121-123, 139
LONELINESS: 145
—marriage-infidelity, 220
—sex, 285-286
LOVE: 292-297
—and becoming, 295
—and communication, 295
—crises of, 295
—and decision-making, 295
—errors about, 293
—image of, 283
—and openness, 296
—and student opinions of, 297-298

McFAUM, JOHN & EDITH, 68-70, 74
MAN AND WOMAN: 159-195
—anti woman prejudices, 163-167
—and dialog, 160-163

– differences and roles in dating, 167-168
– during dating, 169-174
– masculine and feminine, meaning of, 180-184
– open relationship, 160-163
– physical, 169-170
– physically weaker than men? 177-180
– psychological, 172-174
– roles in marriage, 181-184
MARRIAGE:
– and advice, 299
– alleged causes of divorce, 199-200
– communication, 191
– decision-making, 191, 248-249, 285-286, 301
– decision to marry, 270-303
– dialog, 279, 286-287
– dialogic relationship, 191-195, 286-287
– early, 254
– economic unit, 277
– failures (divorce), 197-251
– feelings before decision to marry, 291, 292, 299
– fidelity, 279, 280
– ideal age for, 257-262
– image of, 191-195, 220, 281-284
– and immaturity, 239, 254
– legal age for (chart), 256-257
– and maturity, 258
– nature of mature marriage, 275, 301-303
– new social unit, 290
– patterns for selecting partner, 28
– permanency, 280-281
– sacrament, 277-279
– sexual relationship, 284-285, 287
– student opinions of decision to marry, 271-274
MARRIAGE COUNSELORS, 299
MATURITY: 267-269, 142, 173
– in marriage, 258
MENTAL ILLNESS: 245-246
– and divorce, 199-200
MONEY:
– decision-making, 249-250
– and divorce, 199-200
– and marriage decision, 277, 301
– problems in marriage, 248-250
MOTHER, 40, 41, 42, 45, 82, 143-144, 180
– unmarried, 146, 149, 151

NEARNESS, 44-45, 143-144, 232
NECKING, 135, 145-146, 148-153

OPENNESS, 121-125, 163

PARENTS, 27, 32, 41, 61, 78-79, 82, 88, 97-101, 108-115, 135, 151, 156-158, 185, 230, 239
PEREIRA, ALBERT F., 73-74
POPULATION, 253-254
PREJUDICE, 124-125, 141, 163-167
PROMISES, 231-233

QUOTATIONS:
Alexander, VII – 226; Anacreon – 165; Aquinas – 165; Aristotle – 165; Augustine, St. – 296; Augustus, Caesar – 225;
Batcheller – 227; Bhartrihari – 166; Blackstone – 167; Bordleau, Nicholas – 76;
Cabell – 166, Charcot – 125; Chesterfield – 76, 167;
Decasseres – 165; Dumas – 165;
Epictetus – 66; Ethelbert – 229;
Fields, W.C. – 201, 205;
Genesis – 221; Greene, Graham – 281;
Hammurabi – 224; Hazlitt – 125; Herbert, George – 180; Hunt, Leigh – 76;
Ingalls – 166;
Jaspers – 65; Johnson, Sam – 77; Justinian – 225;
Lawrence – 129; Leeuwenhoek, Anton van – 179; Lex Julia de Adulteriis – 225;
Magnani – 65; Manu – 167, 216; Matthew – 216; Montesquieu – 225;
Proverbs – 76, 277, 292;
Sandford, James – 295; Schopenhauer – 166; Shakespeare – 180; Stanton – 65; Sullivan, F. – 71;
Talmud – 225; Taylor, Jeremy – 218;
Wilde – 167;
Yeats, J.B. – 280.

REBELLION, 98-101, 253-257
REJECTION:
– calling off a date, 114-115
– communication, 86
– dialog, 123
– intercourse, 242-245
– non-listener, 87
– parents, 61-62, 91
– vs. becoming, 39-43
RELIGIOUS DIFFERENCE, 247-248
– and divorce, 199-200

SEEMING, 120-121, 293
SELF:
– actual self, 46-48, 86, 126, 142, 269, 295
– and becoming, 48-49
– communication of, 85-86
– decision-making, 81-83
– images, 53-63, 102, 141, 148, 166, 209-210
– and love, 297
– potential self, 46-48, 86, 126, 142, 268
– true self, 47, 119-120
SEX-CENTRIC DATE, 145-152
– consequences of 149-152
SEX DIFFERENCES, 162-163
– primary and secondary, 169-175
SEXUAL INCOMPATIBILITY, 240-245
– and adultery, 220
– and divorce, 199-200
SHETTLES, LANDRUM B., 179
STEADY DATING:
– advantages of, 263
– breaking off, 132-134
– decision-making, 264-267
– fidelity, 265
– maturity, 264
– personal becoming, 267
– student opinions of, 264-265
– teenage marriages, 263-266
SUGERMAN, DANIEL, 149

TEMPERAMENTAL CONFLICTS, 89, 237-238
– and divorce, 199-200
THOMAS, JOHN L., 198-200, 201-203, 222, 234-250, 274
TORS, IVAN, 144

VIRGINITY, 152, 154-156

WEDDING, 275
WOMAN: 175-176
– creation of, 159-160
– growth and development, 175-176
– menstruation, 175-176
– suffrage, 33
– weaker sex?, 177-180
– work (chart), 37